Springer Aerospace Technology

The series explores the technology and the science related to the aircraft and spacecraft including concept, design, assembly, control and maintenance. The topics cover aircraft, missiles, space vehicles, aircraft engines and propulsion units. The volumes of the series present the fundamentals, the applications and the advances in all the fields related to aerospace engineering, including:

- structural analysis,
- aerodynamics,
- aeroelasticity,
- aeroacoustics,
- flight mechanics and dynamics,
- orbital maneuvers,
- avionics,
- systems design,
- materials technology,
- launch technology,
- payload and satellite technology,
- space industry, medicine and biology.

The series' scope includes monographs, professional books, advanced textbooks, as well as selected contributions from specialized conferences and workshops.

The volumes of the series are single-blind peer-reviewed.

To submit a proposal or request further information, please contact: Mr. Pierpaolo Riva at pierpaolo.riva@springer.com (Europe and Americas) Mr. Mengchu Huang at mengchu.huang@springer.com (China)

The series is indexed in Scopus and Compendex

Alexander Nikolaevich Akimov ·
Vadim Vadimovich Vorobyov ·
Dmitry Alexandrovich Zatuchny

Limiting Modes of Aircraft Flight

Methods, Algorithms and Results

Springer

Alexander Nikolaevich Akimov
Moscow State Technical University of Civil Aviation
Moscow, Russia

Vadim Vadimovich Vorobyov
Moscow State Technical University of Civil Aviation
Moscow, Russia

Dmitry Alexandrovich Zatuchny
Moscow State Technical University of Civil Aviation
Moscow, Russia

ISSN 1869-1730 ISSN 1869-1749 (electronic)
Springer Aerospace Technology
ISBN 978-981-19-6328-5 ISBN 978-981-19-6329-2 (eBook)
https://doi.org/10.1007/978-981-19-6329-2

This Springer imprint is published by the registered company Springer Nature Singapore Pte Ltd.
The registered company address is: 152 Beach Road, #21-01/04 Gateway East, Singapore 189721, Singapore

Introduction

Analysis of the accident rate of aircraft shows that a significant number of accidents (more than 70%) occur on technically serviceable aircraft due to erroneous actions of the crew. This is primarily due to the tendency to complicate the control processes of modern aircraft. Moreover, this complication has already approached the limit of the psychophysiological capabilities of a person. The situation is aggravated by the need for the pilot (crew) to remember a large number of restrictions that outline the area of operational flight modes.

As a result, more than 50% of accidents with technically serviceable aircraft occur due to the release of the aircraft to the limit flight modes and collisions with the ground surface in controlled flight. To solve the problem of accidents by improving only the training of the crew is quite problematic due to the high psychophysiological tension of the crew, when flying near restrictions, caused primarily by the peculiarities of the external environment and ergonomic shortcomings of aviation equipment.

Onboard the aircraft, there are special systems that prevent it from reaching the maximum flight modes. However, they do not cover the full range of restrictions imposed. Their disadvantages are mainly related to the low accuracy of maintaining the limit boundary, the possibility of casting according to the limited parameters, the inability to simultaneously limit several parameters, and the protracted transients. In addition, the existing systems for maintaining constraints are developed on different theoretical approaches, with varying degrees of heuristics.

Many works have been devoted to the problem of developing a unified theoretical basis for creating systems for maintaining restrictions. However, it should be noted that, despite the severity of the problem of ensuring flight safety, at present, this direction is undeservedly forgotten.

Summarizing the above, we can conclude that to solve the problems of maintaining restrictions on the parameters of the movement of an aircraft (AC), controlled by a not always optimal crew, no universal method has been developed at present.

The book discusses the method of maintaining constraints on the components of the state vector of a dynamic system. The idea of the method is to analyze such an integral indicator as the distance in metric space to the constraint surface, given in some way. The developed method provides the following:

- obtaining solutions to various problems of maintaining the specified constraints on the components of the state vector of a dynamic system with different methods of setting its mathematical model;
- obtaining a realizable trajectory of withdrawal in phase and physical space from the constraint surface with a given constraint control;
- adaptability to the properties of the object and the shape of the constraint surface.

Contents

1 Onboard Restraint Systems. State of the Issue. Formulation of the Problem 1
1.1 Aircraft Accident Analysis 1
1.2 Operational and Limiting Flight Modes 3
1.3 The Main Types of Restrictions When Piloting an Aircraft 4
1.4 The Psychological Characteristics of the Pilot When Flying Close Restrictions 6
1.5 Classification of Onboard Systems for Limiting the Limiting Flight Modes. Typical Algorithms for Their Functioning 9
1.6 Analysis of Existing Methods for the Synthesis of Systems for Maintaining Constraints. Research Problem Statement 15
References 16

2 The Method of Adaptive Maintenance of Constraints on the Components of the State Vector of a Dynamic System 19
2.1 Substantial Problem Statement 19
2.2 General Case of Keeping the Constraint 20
2.3 Proof of the Analogy of the Method of Maintaining Constraints to the Fundamental Law of Equal-Variable Motion 25
2.4 Synthesis of Algorithmic Support for an Adaptive Constraint System for Particular Models of a Dynamic System 28
2.4.1 Some Ways to Describe Stationary Constraint Surfaces 28
2.4.2 Algorithm for Constraining a Deterministic Linear Continuous Dynamic System 31
2.4.3 An Algorithm for Constraining a Deterministic Continuous Controlled Multilevel Dynamical System 32
References 34

3 Algorithms for Adaptive Limitation of Aircraft Flight Parameters 35
3.1 Adaptive Limitation of Aircraft Angles of Attack 35
3.2 Peculiarities of Maintaining Restrictions on the Observed Parameters of Aircraft Movement 41
3.3 Limiting the Roll Angle of the Helicopter 46
3.4 Restriction in Space of Two Components of the State Vector by the Example of the Problem of Limiting the Angles of Attack and Slip of an Aircraft 50
References 54

4 Algorithms for Adaptive Limitation of Trajectory Parameters of Aircraft Movement 55
4.1 General Case of Limiting the Trajectory of Aircraft Movement in Three-Dimensional State Space 55
4.1.1 Substantial Problem Statement 55
4.1.2 Object Model and Constraint Surfaces 55
4.1.3 Derivation of the Calculated Dependencies of the Limiting Algorithm 57
4.2 Particular Problems of Limiting the Trajectory Parameters of Movement in the Vertical and Horizontal Planes 59
4.2.1 Aircraft Drift in the Vertical Plane 59
4.2.2 Aircraft Drift in the Horizontal Plane 65
4.3 Synthesis of Algorithms for Maintaining Limitations Within the Range of Altitudes and Flight Speeds 68
4.4 Features of Maintaining Non-Stationary Constraints 71
References 73

5 Aircraft Drift Away from Limiting Surfaces Along Programmed Trajectories 75
5.1 Substantial Problem Statement 75
5.2 Algorithms for Calculating the Angles of the Aircraft Position in Space in the Case Movement It Away from an Obstacle 76
5.3 Conditions for the Activation of «Withdrawal» Mode 80
5.4 Formation of Controls 82
5.5 Solving Private Problems of Aircraft Evacuation from Obstacles 83
5.5.1 Vertical Wall 83
5.5.2 Inclined Plane 86
5.5.3 Comparative Analysis of Algorithm «Withdrawal» 87
References 91

6 Method and Algorithms for Direct Optimization of the Movement of a Damaged Aircraft 93
6.1 Analysis of the Main Prerequisites for Creating a Method of Direct Optimization 93
6.2 A Loose Proof of the Correspondence of the Principle of Variation of the Derivative Function of an Arbitrary Curve to the Necessary Optimality Condition 95
6.3 Development of the Main Provisions of the Method of Direct Optimization of the Motion of Dynamic Systems 99
6.3.1 General Formulation of Optimal Control Problems and Determination of the Sequence of their Solution in the Method of Direct Optimization of Dynamic Systems 99
6.3.2 Level-by-Level Decomposition of the Mathematical Model of the Control Object 103
6.3.3 The Main Provisions of the Method of Direct Optimization 105
6.3.4 Organization of the Procedure for Maintaining the Restrictions on Coordinates of State and Control Vectors 120
6.3.5 The Procedure for Optimizing the Basic Trajectory Movement of Aircraft 130
6.4 Estimation of the Reliability of the Method of Direct Optimization Using Examples of Solving the Simplest Optimal Control Problems 135
6.5 Synthesis of the Laws of Control of the Pilot Circuit Based on the Method of Direct Optimization 144
6.6 Technique and Features of Aircraft Thrust Vector Control 156
6.6.1 Aircraft Thrust Vector Control Algorithm 158
6.6.2 Analysis of the Performance of the Thrust Vector Control Algorithm 161
6.7 Extended Version of the Method of Direct Optimization 167
References 172

Abbreviations

AA	Air accident
AC	An aircraft
ACS	Automatic control system
ADOR	Analytical design of optimal regulators
AE	Aviation equipment
CCV	Classical calculation of variations
CS	Constraint surface
DS	Dynamic system
LML	Limit mode limiter
MFM	Manual for flight maintenance
OBDCS	Onboard digital computing system
RTV	Rotary thrust vector
SLAA	The system for limits of the angle of attack
SLS	System of limiting signals
VOF	The violations in the operations of flights
MDO	Method of direct optimization
MM	Mathematical model
OC	Optimal control
OL	Overload limiter

Chapter 1
Onboard Restraint Systems. State of the Issue. Formulation of the Problem

1.1 Aircraft Accident Analysis

Ensuring flight safety is one of the most important problems in aviation. Flight safety is determined by a large number of different factors. The ranking of these factors by their contribution to the accident rate, as well as the assessment of the accident rate itself, can be most reliably carried out, using statistical methods. Since aviation accidents are relatively rare events, a sufficient statistical sample can be obtained only with a flight of about 1.5 … 2 million hours. This is roughly equivalent to an annual flyover.

Each accident is a unique event, so statistical analysis is possible only for generalized groups of causes. Thus, from the total number of aviation accidents, it seems appropriate to single out a group of causes, caused by unpredictable factors, including the perceived risk and the psychophysiological characteristics of a person as an operator. The remaining aviation accidents (AA), in principle, should be prevented by preventive methods, but in fact more than 70% of AA falls on them.

Of particular concern are accidents, involving technically operable aircraft (AC) (Fig. 1.1). The number of aviation accidents, associated with erroneous actions of the crew on operable equipment, currently amounts to more than 70% and has a growing trend. Many failures of aviation equipment do not inevitably lead to an accident. However, in many cases, incorrect actions of the crew aggravate the consequences of failure and add to the statistics of aviation accidents (Fig. 1.2).

If we analyze the accident rate by the factors of causes, then the largest number of AC occurs due to violations (omissions) in the operations of flights (VOF). At the base of these accidents is most often the exit of the aircraft beyond the limit of flight modes, namely the limits on the angles of attack and roll, vertical overload, altitude, and speed of flight are exceeded. In addition, a significant number of aviation accidents (more than 20% of their total number) are associated with the collision of aircraft with the earth's surface.

Of particular concern is the fact that a collision with the terrain in most cases ends in disasters. Thus, the accident severity coefficient, which is the ratio of the number

V. Akimov and V. A. Zatuchny, *Limiting Modes of Aircraft Flight*,
Springer Aerospace Technology, https://doi.org/10.1007/978-981-19-6329-2_1

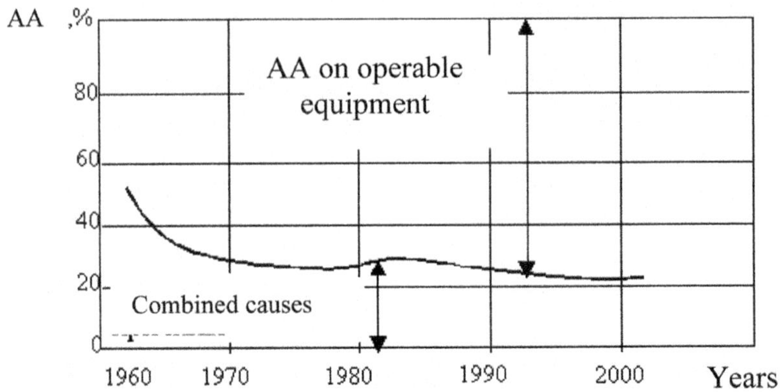

Fig. 1.1 Analysis of aviation accidents with technically operable aircraft

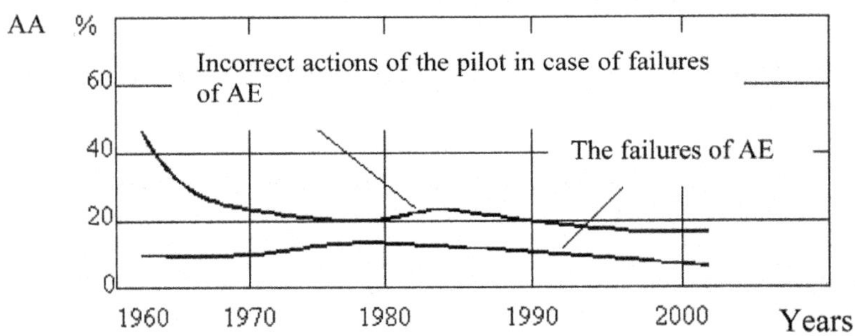

Fig. 1.2 Effects of improper actions of the crew on the statistics of accidents

of accidents during a certain period of operation to the total number of accidents during the same period of operation, is over 0.85 [1, 2].

A sufficiently high proportion of AA, caused by the collision of an aircraft with the earth's surface, can be explained by the following reasons [2]:

- the tendency of pilots, especially highly qualified ones, to overestimate their capabilities and disrupt flight tasks;
- insufficient effectiveness of onboard technical means to prevent the aircraft from colliding with the terrain;
- the decline in the practical training of the crew in recent years.

  Analyzing the circumstances and causes of the AA that led to the collision with the earth's surface, it should be concluded that the main mistakes of the crew are the following errors and shortcomings in flight work:
- it is not set, or the value of the safe altitude on the radio altimeter is set incorrectly before departure, or the crew for some reason does not respond to the alarms;
- maintaining the set height, errors in reading the height, or loss of height control during a dive during aiming [3];

- errors of the crew, when removing the aircraft from a difficult position after stalling or loss of spatial orientation.

Thus, assessing the state of flight safety, the human factor came out in the first place in terms of the contribution to the accident rate, and a significant number of aviation accidents for this reason are associated with the withdrawal of the aircraft beyond the limit flight modes. All this allows us to conclude that the problem of preventing the aircraft from going beyond the limits is one of the most important problems in ensuring flight safety.

1.2 Operational and Limiting Flight Modes

Aircraft can fly only in a certain range of kinematic parameters (speeds, altitudes, angles of position of the aircraft, the speed of their change, etc.) and parameters that determine the operation of the functional systems of the aircraft. Going out of this range leads to the occurrence of an emergency (or catastrophic) situation and to the threat of an adverse event, such as an aircraft accident. Formally, we can assume that in the phase space of these parameters, there is a surface that divides this space into two areas: operational and critical modes [4].

This surface, as you know, is called critical. However, in fact, taking into account the existing uncertainty in the description of the atmosphere (turbulence, precipitation, etc.), minor differences in the characteristics of aircraft, and the peculiarities of piloting by pilots, this surface turns into a layer of a certain thickness. For safety reasons, in this case, the critical surface must be taken as the side of the layer that is located on the side of the operational area. However, the pilot is not able to accurately maintain the trajectory in the considered phase space.

To prevent the aircraft from leaving the critical surface, a new surface is introduced, which further narrows the operational area. The resulting area is called the area of acceptable flight modes. Approximating the critical surface with a hyperplane, it is possible to set limits on isolated parameters, rather than on their functions, as the boundary of the domain of acceptable modes. This is absolutely necessary if the control over the maintenance of restrictions is carried out by the pilot, who must remember all the values of the restriction parameters, specified in the Manual for Flight Maintenance (MFM) of the AC type.

The purpose of this book is to automate the prevention of parameters going beyond the assigned limits. As constraints, their values are accepted, which are set in the MFM of the aircraft type.

1.3 The Main Types of Restrictions When Piloting an Aircraft

The reasons that make it necessary to impose restrictions on the parameters of piloting an aircraft (including its trajectory in physical space) are as follows:

- features of aerodynamics;
- strength characteristics of the aircraft structure;
- unacceptable intersection of the trajectory of the aircraft with the relief of the underlying surface;
- combinations of the above reasons.

Features of aerodynamics, in turn, are associated with a change in the nature of the flow around the aircraft at high angles of attack and glide, the limit numbers of flight M, as well as during the transition from subsonic flight speed to supersonic and back. To avoid falling into the area of unfavorable flow, the maximum angle of attack and the number of M of flight are limited. Figure 1.3 shows the ranges of operational, as well as intermediate non-operational (shaded in the figure) and supercritical angles of attack. In addition, the maximum α_{lim}, α^*_{lim} and critical α_{cr}, α^*_{cr} angles of attack are indicated (asterisks indicate parameters, related to flying at negative angles). Reaching the supercritical angles of attack $\alpha > \alpha_{\text{cr}}$ and $\alpha < \alpha^*_{\text{cr}}$ is dangerous by stalling the aircraft.

It should be noted that the presence of sliding reduces the range of operational angles of attack. Access to large angles of attack is possible not only in aerobatics, but also at low flight speeds when the lifting force at subcritical angles of attack is insufficient to balance the weight of the aircraft. Therefore, for a steady rectilinear horizontal flight without roll and slip, the minimum permissible flight speed ($V_{min\text{lim}}$)

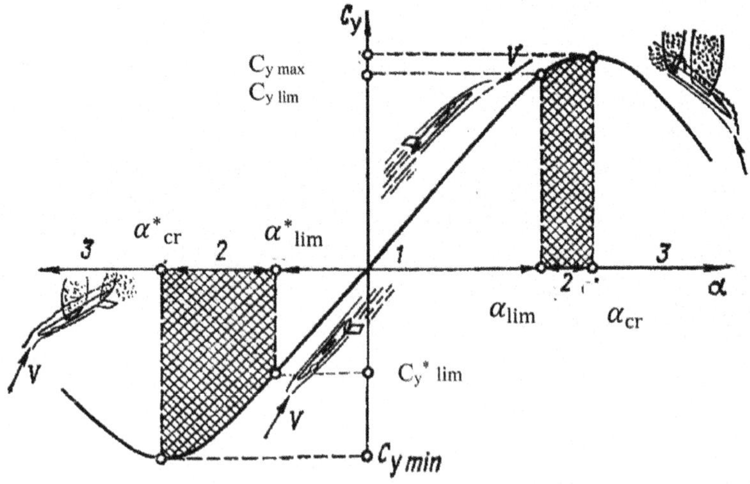

Fig. 1.3 Ranges of operational, intermediate non-operational, and supercritical angles of attack

is taken as the minimum speed limit. It is defined as the speed of flight at a given altitude at the maximum permissible values of the lift coefficient $C_{y_{\lim}}$ (angle of attack a $\alpha_{\lim}$) in normal flight operation of the aircraft. Going to $V_{min\lim}$ is also dangerous to the occurrence of spontaneous vibrations of the aircraft in pitch, roll, and yaw.

The maximum limit number $M_{\lim}$ is the maximum number of flight M, allowed in normal flight operation at a given altitude, given the flight configuration, engine operating mode, and flight weight. Exceeding the $M_{\lim}$ in flight is dangerous, first of all, by the appearance of undesirable features in the behavior and piloting of the aircraft, for example, loss of track stability. In heavy and limited maneuverable aircraft, the $M_{\lim}$ value can be associated with the occurrence of a wave crisis, the appearance of dangerous features in behavior, and piloting at transonic flight M numbers.

The strength characteristics of the AC structure impose restrictions on the maximum lifting force (the threat of breakage or deformation of the bearing surfaces of the aircraft) and on the speed pressure. The maximum overload n_y and the maximum instrument speed $V_{\lim}$ are physically restricted. In addition to limiting positive overload values, the MFM also sets negative overload values. However, in practice, access to them is extremely rare. In addition to these limitations, there may be other reasons for limiting normal overload. The positive overload limit may depend, for example, on whether the flight is performed with or without an anti-overload suit. Negative normal overload is limited by the physiological capabilities of the crew, the operating conditions of the fuel system, and the like.

For heavy aircraft, as a rule, a limit on the maximum roll is set. The dangerous consequences of exceeding the roll angle at low altitudes are the loss of altitude, when the aircraft is banked, and at high altitudes—dragging it into a steep spiral. The reasons that contribute to the possibility of involuntary roll the plane quite a lot, for example, failure of the engine, located on one of the halves of the wing, uncontained failure of automatic control system (ACS) on the roll, aileron reversal, the aircraft wake turbulence trail, the reverse reaction roll to the control actions of the pilot, etc. However, it is obvious that for any aircraft under consideration, it is impossible to name one limit value of the roll angle, exceeding which is dangerous. Depending on the flight conditions, primarily on the flight altitude and speed, this limit changes significantly. The lower the flight altitude at the time under consideration and the closer the flight speed is to the minimum allowed speed for a given AC, the lower the roll angle can be considered acceptable.

A more difficult task of restriction is the task of preventing the helicopter from reaching an unacceptable roll angle [5]. This is due to the fact that, firstly, the allowed roll angle depends significantly on the flight altitude (usually one degree of roll per 1 m of height), and, secondly, to eliminate the roll on a helicopter (unlike an airplane), you have to use almost all the controls.

A very complex type of restriction is acquired when piloting at low altitudes and when individual heights of the underlying terrain exceed (or are close in height) the geometric altitude of the flight. Piloting in these conditions should guarantee the implementation of a flight path that does not intersect with the terrain anywhere.

From the above, it can be seen that the parameters of the movement of a modern aircraft are subject to numerous restrictions, which vary depending on the flight conditions, the position of the AC in space, its configuration, and other factors. However, the task of preventing going beyond the established limits is currently assigned mainly only to the pilot. Although there are special technical devices to help him onboard the aircraft, or there are appropriate modes of operation of the automatic control system, but, judging by the above statistics, the effectiveness of the actions of the crew and automatic devices is low. So, for example,

- on modern aircraft, the value of $V_{min\lim}$ is only indirectly limited by the automatic angle of attack limitation. In some cases, there is a passive alarm about the exit to $V_{min\lim}$, informing the pilot about the fact of entering the stall modes;
- there is no automatic restriction of $V_{\lim}$ and $M_{\lim}$ on aircraft. However, no accidents were registered during the analyzed period due to the departure of aircraft for the specified parameters. Therefore, the problem of going beyond $V_{\lim}$ and $M_{\lim}$ is not as acute as going to $V_{min\lim}$;
- to prevent collisions with terrain, special systems are installed on aircraft, or standard ACS modes are used. However, they are either passive devices that only inform the crew about an unacceptable approach to the ground, or they are turned on only when flying in automatic mode;
- automation of the roll angle limitation is carried out mainly on heavy aircraft and is performed in the form of passive signaling of the maximum roll angle. The pilot's omission of a signal about reaching the maximum angle of roll with the subsequent development of an emergency situation is a fairly frequent event.

1.4 The Psychological Characteristics of the Pilot When Flying Close Restrictions

The limits set in the MFM, as already noted, do not coincide with the critical values of the parameters, exceeding which directly leads to an emergency or catastrophic situation. However, exceeding this or that limit for the pilot corresponds to getting into a special situation and is accompanied by a complication in his work.

In general, a special situation is very diversified in terms of impacts on the functioning of the circuit, «the crew–AC», and in its effects on the human psyche, that is, the level of informativeness of those signals, which appear to the pilot at the time of occurrence of the particular situation.

A special situation causes two levels of pilot responses: adaptive–protective reactions (reflex) and complex intellectual actions, related to assessing the situation, ensuring the process of distributing attention between monitoring the current state of the AC and its systems and developing a new, unplanned, action plan, and predicting the development of a special situation.

A special situation by its nature is a psychogenic effect. It is addressed mainly to the human psyche and depends on his mental state, that is, according to its content, a special situation can be perceived quite differently by different pilots.

The pilot contrasts the impact of a special situation with mental readiness, which consists of two components:

- mental stability, due to the state of the body;
- mental stability, due to professional training and the general level of mental qualities of the individual.

When controlling the AC and its systems, the pilot, along with the main functions for the implementation of the flight task, performs the role of an ergatic reserve to ensure flight safety. Its activity in the event of a special situation is generally reduced to four consistently implemented stages.

In the first stage, in a special situation, the following features occur in the search and perception of information:

- the transition from quantitative reading of instrument information to qualitative reading (to speed up the search for a dangerous signal);
- skipping useful signals due to increased concentration of attention on other objects;
- there may be distortions in the perception of useful signals, which are characterized either by the perception of the expected signal instead of the real one, when the situation is incorrectly assessed, or by an incorrect assessment of the perceived signal due to the ambiguity of the situation.

The success of the perception of information will depend, in particular, on the extent to which the incoming signals perform the following functions:

- draw attention to the special situation that actually occurred;
- provide an understanding of the causes of the incident;
- will help to eliminate erroneous actions, when moving from the current activity to parrying a special situation.

Receiving information about a particular situation begins with drawing attention to a new stimulus. The alarm usually appears unexpectedly. The pilot at the moment of receiving the signal is busy with work, which requires visual and auditory control and absorbs his attention. Therefore, devices must be installed onboard the AC to ensure a high attractive effect in the event of a special situation.

In some cases, the appearance of an alarm signal before the perception of the displayed parameters and comparing them with the nominal values causes the onset of a special mental state—a conflict of ideas. This state is accompanied by impulsive motor reactions, various kinds of tentative movements, etc., which delay the decision-making time.

In the second stage, the elements of analysis of the actual information received from the control object already prevail in the operator's activity. At this stage, it is being implemented:

- the selection of operator the nominal values of the monitored parameters from the memory X_{i_n}, which must be carried out in a time interval $\tau < \tau_{\lim}$;
- comparison of actual X_i and nominal values X_{i_n} of monitored parameters by the definition of the error $X_i - X_{i_n} \leq \Delta X_{i_{\lim}}$ and its sign $\text{sign}\Delta X_{i_{\lim}}$;
- estimation of the rate of change of the controlled parameters $\dot{X}_i$, $\dot{X}_{i_n}$ and their deviations from the nominal values $\Delta \dot{X}_{i_{\lim}}$.

The complexity of the operator's work at this stage is determined by the number of controlled parameters, according to the totality of the values of which a conclusion is made about the danger of the situation.

$$S = \{\Delta X, \Delta \dot{X}, \text{sign}\Delta X\}. \tag{1.1}$$

In practice, such procedures as timely and correct selection of the necessary information from memory that characterizes the nominal management process and assessment of the current situation, implemented on the basis of comparing a set of features with the corresponding area of their limitations, are often difficult to implement.

In the third stage, the operator makes a decision on the nature of the control action U_i, which must be implemented in the current situation. This stage can be formally described as follows:

$$\begin{aligned} &U_1 \quad \text{if} \quad S = S_1\{\Delta X,\ \Delta \dot{X},\ \text{sign}\,\Delta X\}; \\ &U_2 \quad \text{if} \quad S = S_2\{\Delta X,\ \Delta \dot{X},\ \text{sign}\,\Delta X\}; \\ &\ldots\ldots\ldots\ldots\ldots\ldots\ldots\ldots\ldots\ldots \\ &U_m \quad \text{if} \quad S = S_m\{\Delta X,\ \Delta \dot{X},\ \text{sign}\,\Delta X\}, \end{aligned} \tag{1.2}$$

that is, for each state of the object S_i, the operator must have an appropriate solution U_i, $i = 1, \ldots, m$, representing the sequence of its actions to localize a special situation.

The difficulty of this stage is the need to remember a large number of pairs $S - U$ (state–decision) and limited time to make a decision in special situations.

When an unexpected condition occurs S_{m+1}, the operator is experiencing serious difficulties in making a decision. In this case, he carries out a sequence of trial actions, choosing the most acceptable, and often does nothing, or makes the wrong decision.

In the fourth stage, a control action or a set of control actions is implemented. This stage is the final one, and in case of its erroneous or untimely implementation, all the previous activity of the operator is «nullified».

The difficulty of implementing this stage is that the operator may not find the necessary switch, button, or control lever. Do not be able to touch it, when exposed to alternating overloads and angular accelerations. It is a mistake to act on another, nearby lever or switch. Do not balance the forces and movements of the control levers with the required ones. The difficulties that arise are compounded by the stressful state of a person in a special situation.

As a result, the success of the withdrawal from a special situation depends not only on the training of the crew, but also, first of all, on their psychological preparation and readiness to act in a special situation.

Based on the above, it is desirable to have a flight safety system onboard the AC, the task of which is to strictly limit the operational range of flights in all parameters that have a direct impact on flight safety [6–8]. Such a system should perform the following functions:

- issue signals with a high attractive effect, in order to timely notify the AC crew about the occurrence of a special situation;
- provide an accurate assessment of the actual values of the monitored parameters, storage of their nominal values, and comparison of actual values and nominal values. Such a task can be implemented by onboard computers, which have a large memory and high performance;
- determine a point in the state space for the active intervention of the system in the control in case of inactivity or incorrect actions of the pilot. The position of this point should be adjusted depending on the confidence in the control system, the state of the AC, external conditions, and other factors;
- to perform a given sequence of control actions.

1.5 Classification of Onboard Systems for Limiting the Limiting Flight Modes. Typical Algorithms for Their Functioning

Onboard systems for limiting flight limits are one of the most important components of a comprehensive flight safety system [9].

Existing restriction systems can be classified into several levels.

According to the level of activity, they are divided into passive, active, and semi-active systems. Passive systems include systems that only inform the crew about the approach of the determining parameter (a parameter that directly affects the safety of the flight) to its limit value.

An example of them can be various types of signaling devices and pointers, for example an indicator of the angles of attack and overload, a signal board, and other devices.

Active systems provide automatic warning of AC output beyond the flight limits. These include the so-called automatic devices, which limit mode limiting and are represented, depending on the type of AC, by a system for limiting the angle of attack and (or) overload, the ACS mode "Withdrawal from a dangerous height" (provided that it is previously activated).

Semi-active systems provide information to the crew and form recommendations to the pilot to prevent special situations, associated with exceeding the limit parameters in flight. These systems are most widely represented on the fourth-generation AC.

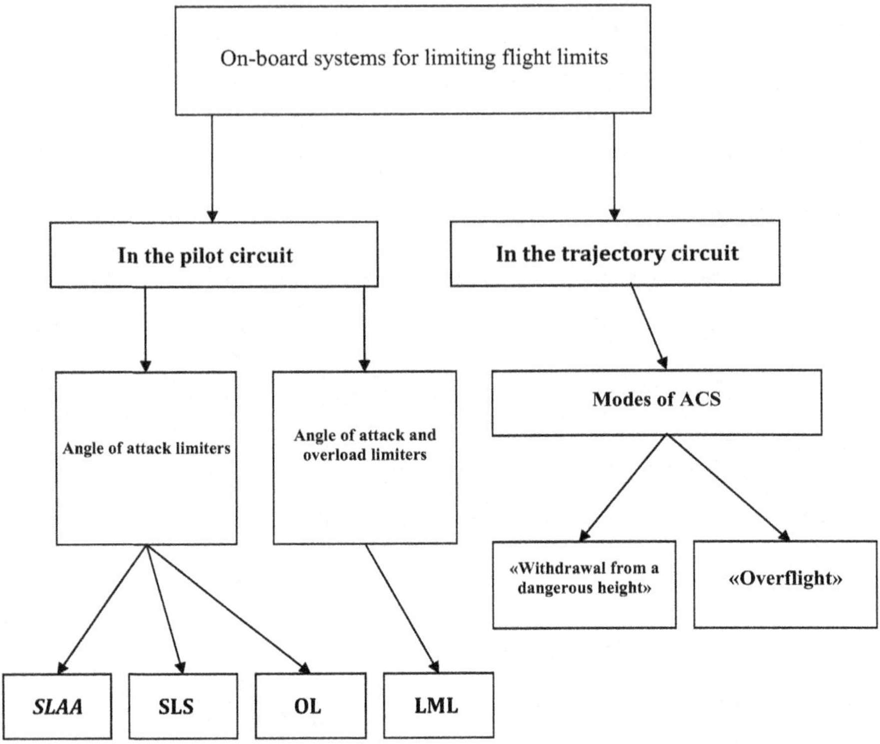

Fig. 1.4 Classification of existing systems for limiting the impact on the control circuit of the aircraft

The classification of existing systems of restriction on the impact on the control circuit of the aircraft, in relation to active systems, is shown in Fig. 1.4. It is marked: SLAA—the system for limits of the angle of attack, SLS—the system of limiting signals, OL—overload limiter, and LML—limit mode limiter.

Despite a large number of restrictions on the flight parameters, the automation of maintaining the restrictions applies only to the angles of attack, overload, and minimum flight altitude. On helicopters, there is no automation, related to flight safety improvement systems.

The operation of a fairly wide range of onboard systems to withstand restrictions is essentially the following. If there is no threat to flight safety, then the restriction system does not manifest itself in any way, and its output signals are identically equal to zero. As soon as a threat to flight safety appears or is predicted, associated with the release of the determining parameters of the movement beyond their limit values, output signals are generated, aimed at removing the threat to flight safety.

The formal record of the functioning algorithm is reduced to the following form:

$$\boldsymbol{U} = \begin{cases} 0, \ if \ X \in G; \\ \boldsymbol{U^*}, \ if \ X \notin G. \end{cases} \tag{1.3}$$

Here U is understood as the output of the system constraints; U^*—a specific sequence of actions (lighting system, an information message, shaking the handle control, repel control knobs, and the like); O—controlled by restraint system the state vector AC; G—a certain range of acceptable values of the motion parameters (the vector component O).

Expression (1.3) reads as follows. Signal $U = 0$, if X is inside the allowed area G. Signal $U = U^*$, if X is out of the allowed range. Often, as X in (1.3), its predicted value is used in one way or another. The most common method for predicting the state vector is linear extrapolation. Since, in most cases, there are no meters of the derivatives of the controlled parameters onboard the AC, various schemes of their differentiation are used.

Most of the constraint systems are built according to the algorithm, described above. However, almost all of them have certain disadvantages. To identify these disadvantages, consider the algorithm of functioning the system for limits of the angle of attack (SLAA).

Most of the algorithms of SLAA functioning are reduced to a typical scheme:

$$\delta_e = \begin{cases} \delta_e^p + \delta_e^{\text{SSC}}, & \text{if } \alpha_\text{p} < \alpha_\text{lim}; \\ \delta_e^\text{SLAA} + \delta_e^\text{SSC}, & \text{if } \alpha_\text{p} < \alpha_\text{lim}, \end{cases} \tag{1.4}$$

where δ_e^p—the deviation of the elevator, formed by the pilot during piloting; δ_e^{SSC}—the deviation of the elevator, generated by the system to improve stability and controllability; δ_e^SLAA—the deviation of the elevator, formed SLAA; α_p—predicted angle of attack value; α_lim—the maximum limit value of the angle of attack.

The most subtle point in the SLAA algorithms is the scheme for generating and calculating the predicted angle of attack value. Despite the great variety of schemes for calculating α_p, they can essentially be reduced to a linear extrapolation of the form:

$$\alpha_p = \alpha + K\dot{\alpha}, \tag{1.5}$$

where α—the current value of the angle of attack; $\dot{\alpha}$—the derivative of the change in the angle of attack; K—the coefficient of dynamic lead activation SLAA, that is, the time at which the change in the angle of attack is predicted (extrapolated). As a rule, the K coefficient is either a fixed value or has several program settings depending on the configuration of the aircraft. The current adjustment of the value K depending on the state of the constraint object is not performed. As a result, these systems have the disadvantage that there is no guarantee (and this is essential) of the exact maintenance of the limits at the level of acceptable values.

The control signal from SLAA δ_e^{SLAA} is formed in different ways. On some AC there is a current deviation of the elevator from the pilot at a point in time when $\alpha_p \geq \alpha_{\text{lim}}$, so

$$\delta_e^{\text{SLAA}} = \alpha_e^p \left(\alpha_p \geq \alpha_{\text{lim}}\right) \tag{1.6}$$

Such algorithms are used on AC, which are characterized by a low rate of changes in the angle of attack $\dot{\alpha}$. . At the same time, by limiting the control signal from the pilot, SLAA, together with the system to improve stability and controllability (SSC), prevents a further increase in the angle of attack. Quite often, in the law (1.5), an overload n_y is used instead of a parameter α, that is, the restriction is made according to the normal overload.

On the AC, which are characterized by high rates of change in the angle of attack, other methods of forming δ_e^{SLAA} are used.

Consider two of them.

The first method is that when the value is reached $\alpha_p \geq \alpha_{\text{lim}}$, the stabilizer is gradually moved to a given fixed position. The desired goal of this control is to reduce the rate of increase of the angle of attack to zero at the point =. However, the use of a fixed setting of the dynamic lead coefficient K does not lead to the achievement of the specified goal. As a result, activation of the SLAA either leads to underutilization of the AC capabilities or allows exceeding the limited parameter (show Fig. 1.5).

In Fig. 1.5, the case of the output of the AC to α_{lim}, when the SLAA is activated with different dynamic lead coefficients K, is considered. In the figure, points t_1, t_2, and t_3 determine whether SLAA is enabled.

The second method of formation δ_e^{SLAA} is based on the fact that, starting from the moment, when $\alpha_p \geq \alpha_{\text{lim}}$, further movement of AC to the angle α_{lim} is not carried out by the pilot, but by SLAA, implementing a control law, similar to the astatic ACS:

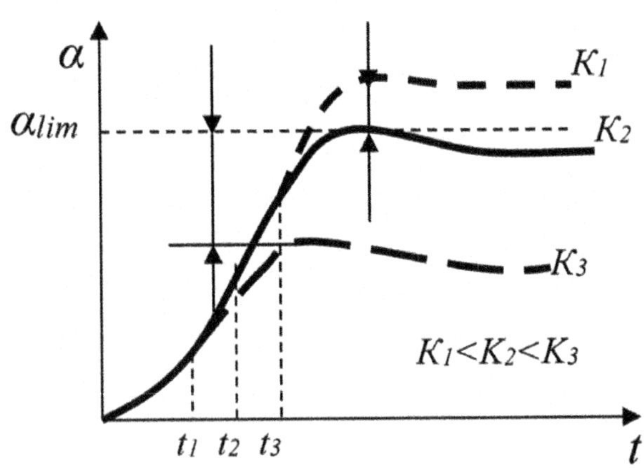

Fig. 1.5 The case of the movement of AC to the maximum limit angle of attack, when the SLAA is activated with different dynamic lead coefficients

$$\delta_e^{\text{SLAA}} = K^{\int} \int_0^t (\alpha - \alpha_{\text{lim}})dt. \tag{1.7}$$

This approach has the advantage that it can ensure that the α_{lim} is not exceeded during piloting. However, the disadvantage of the law (1.7) is that the transition process of the AC to the permissible angle of attack in time is prolonged and the possibility of a loss of stability of the control circuit, which is characteristic of integral laws.

The condition for terminating the SLAA is

$$\alpha_p + \Delta < \alpha_{\text{lim}} \tag{1.8}$$

where Δ—a certain value of the angle of attack that provides hysteresis in activation and «releasing» SLAA.

Another algorithm, the disadvantages of which need to be considered, is the algorithm for the functioning of the system for preventing the collision of AC with the earth's surface.

As a rule, these algorithms are an integral part of the ACS functioning algorithms and are called «withdrawal from a dangerous height».

The algorithm functions according to the following scheme:

$$\delta\_e = \begin{Bmatrix} \delta_e^{pACS}, if\ H_p \geq H_{\text{min}}, \\ \delta_e^{wd},\ if\ H_p < H_{\text{min}} \end{Bmatrix}. \tag{1.9}$$

Here δ_e^{pACS}—control, formed by the pilot or ACS; δ_e^{wd}—control, generated by ACS, operating in the «withdrawal» mode; H_p—the predicted value of the height of flight; H_{min}—minimum permissible flight altitude above the terrain.

This predicted height value is calculated, using the formula:

$$H_. = H_{\text{ra}} + K\dot{H}, \tag{1.10}$$

where H_{ra}—the altitude of the flight over the terrain, measured by a radio altimeter; K—the coefficient of dynamic lead, or the time of predicting the change in altitude; signal $\dot{H}$—rate of change in altitude of flight, and the vertical speed signal V_y is often used as $\dot{H}$, which, generally speaking, is not the same thing in terms of changing terrain elevation.

The control of the AC during withdrawal can be formed in various ways. In some cases, software control is used, which consists of pulse deflection of the stabilizer for carburization by a certain amount, holding it in this position for a period of time τ_{pulse} and then removing the control signal. Graphically, this can be represented in the form, shown in Fig. 1.7. Depending on the flight mode, this control withdrawal provides the transfer of the aircraft to climb of altitude with an angle of inclination of the trajectory 15 … 20°.

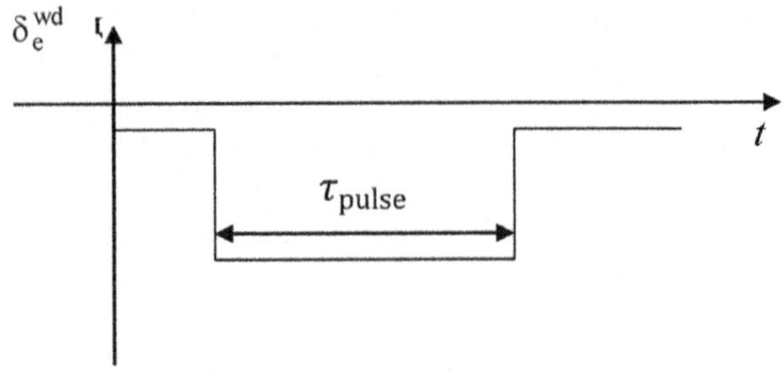

Fig. 1.7 Software control of the aircraft during withdrawal

In other systems, the feedback control is used. In this case, the traditional scheme of constructing the ACS laws is used, which ensures the elimination of the vertical rate of decline of the AC:

$$\delta_e^{wd} = K_{e_1}^{n_y}(n_y - n_{yol}) + K_{e_2}^{n_y}\int_0^t (n_y - n_{yol})dt, \tag{1.11}$$

where the set value of the overload $n_{y_{ol}}$ can be calculated, using the following relations:

$$n_{y_{ol}} = -K_{n_y}^{V_y} V_y + \frac{1 - \cos\gamma}{\cos\gamma} \tag{1.12}$$

or

$$n_{yol} = -K_{n_y}^{\theta}(\theta - \theta_{ol}) + \frac{1 - \cos\gamma}{\cos\gamma}; \theta_{ol} = 8^\circ \tag{1.13}$$

In all cases, the roll angle is eliminated in the side channel. As a result of the consideration of onboard systems for limiting the movement parameters, both in the pilot and in the reactor circuit, the following conclusions can be drawn:

- of all the considered flight limit modes, only the maximum angles of attack and overload are currently limited;
- the main disadvantages of onboard restriction systems are related to the low accuracy of maintaining the border of restrictions, the possibility of casting according to the limited parameters, and the inability to simultaneously limit several parameters.

1.6 Analysis of Existing Methods for the Synthesis of Systems for Maintaining Constraints. Research Problem Statement

The analysis of domestic and foreign literature allows us to conclude that there is a practical lack of systematic scientific research on the synthesis of algorithms for maintaining constraints. At the same time, on the basis of the theory of optimization of processes of management of dynamic objects, there is a theoretical basis for their development. However, despite the fact that the modern theory of optimal control is a fairly perfect and well-developed mathematical apparatus, the problem of the practical application of the developed methods to solving specific problems still remains relevant. In particular, the search for optimal controls in boundary value problems with constraints, as before, largely depends on the experience, intuition, and skill of the researcher. Nevertheless, the most developed algorithms for maintaining the restriction are the algorithms, presented in the work of V.N. Bukov [10]. They are based on the method of analytical design of optimal regulators (ADOR) in the formulation of A. A. Krasovsky. Thanks to the use of predictive models and special penalty functions, they allow us to obtain a high efficiency in maintaining the limitations of short-period motion of the aircraft, which is confirmed by the results, given in the works of A. N. Akimov [11], A. I. Mishchenko, V. D. Gerasimov, and other authors.

However, the main difficulties lie in the synthesis of algorithms for maintaining restrictions in the trajectory circuit, due to the difficulty of obtaining a reliable forecast of the movement of the aircraft at large time intervals. In addition, methods, based on the main provisions of the theory of analytical design, have disadvantages, associated with the presence of subjectivity in the setting of the main parameters, such as the choice of the length of the prediction interval and the matrices of weight coefficients that make up the functional of the generalized work.

Another method that allows us to solve constraint problems is the method of direct optimization, described in the work of A.N. Akimov and V.V. Andreev «The method of direct optimization. The main provisions and the order of application» [12]. In this method, the constraint problem can either be removed from the optimization process and solved independently, like the ADOR method, implementing the functions of the constraint automatic devices, or serve as a mechanism for obtaining optimal realizable trajectories and controls, when solving aircraft control problems.

It is important that these methods do not have the disadvantage, associated with the inability to synthesize control of a dynamic system, in the control circuit in which there is a «purely non-optimal» human operator. This disadvantage is inherent in many aviation tasks when taking into account the direct control of the AC by the crew.

In addition to these, there are a number of heuristic methods and approaches to solving constraint problems [13]. They are presented in the works of G.M. Sienkiewicz, J.I. Shenfinkel, V.F. Smernicki, A.A. Titov, D.G. Ostapenko, and other authors. Their common features are simplifications in describing the constraint surface and the models used. Without suppressing the advantages of heuristic methods, associated with the possibility of practical implementation, their disadvantages include the lack of generality and universality of the approach to the synthesis of constraint systems.

Summarizing the above, we can conclude that no universal method has been developed to solve the problems of maintaining restrictions on the parameters of the AC movement, which is not always controlled by an optimal pilot. Based on the published results [12], the closest approach to solving this problem is the direct optimization method. However, the algorithms for maintaining constraints, developed in accordance with this method, have disadvantages, mainly due to the lack of theoretical elaboration and evidence of the described approaches, as well as the lack of systematic research in various target tasks of constraints.

As a result, to solve the problem of maintaining restrictions on the components of the state vector in the spatial motion of the AC and its movement in phase space, the book suggests the use of a method, which is free from the above disadvantages. The idea of the method is to analyze such an integral indicator as the distance in metric space to the constraint surface, given in some way. The developed method should provide the following:

- obtaining solutions to various problems of maintaining the specified constraints on the components of the state vector of a dynamic system with different methods of setting its mathematical model;
- obtaining a realizable trajectory of withdrawal in phase and physical space from the constraint surface with a given constraint control;
- adaptability to the properties of the object, the shape of the constraint surface, the initial conditions, and the properties of the environment.

References

1. Akimov AN, Vorobyov VV, Lukashov SV (1998) Algorithm for preventing an airplane from colliding with the earth's surface. In: Problems of flight safety. - M.: VINITI, pp 47–50
2. Akimov AN, Vorobyov VV, Lukashov SV (1999) Warning of the collision of the aircraft with the terrain by leading away along spatial trajectories. In: Proceedings of scientific readings, dedicated to the memory of aircraft designer I.I. Sikorsky. Moscow , St. Petersburg, May 25–29
3. Akimov AN, Vorobyov VV, Ferapontov IS (1996) Algorithmic support of the system for improving the safety of low-altitude flight. Ensuring flight safety: Scientific and methodological materials. - M.: Air Force Engineering Academy named by N.E. Zhukovsky
4. Akimov AN, Vorobyov VV (2001) Methods and algorithms for moving the aircraft away from the spatial surface of the constraint. In: Automation and telemechanics. no. 7
5. Pogrebnyak VI (2001) Accident rate in helicopter aviation. In: Problems of flight safety. - M.: VINITI
6. Ukolov IS, Kirsanov BV, Noskov YV (1981) Ensuring flight safety with the help of an active control system/issues of cybernetics. In: Problems of aviation and space cybernetics. - M.: Scientific Council of the USSR Academy of Sciences on the space problem «Cybernetics»
7. Ivanov VS (ed) (2003) Flight safety of aircraft. - M.: Air Force Engineering Academy, named by N.E. Zhukovsky
8. Starikov AN (ed) (1988) Flight safety of aircraft: Methodological foundations. -M.: Transport

9. Denisov VG (1972) Onishchenko V.F. Engineering psychology in aviation and cosmonautics. - Moscow: Mashinostroenie
10. Bukov VN (1987) Adaptive predictive flight control systems. - M.: Nauka
11. Akimov AN (1984) Information interaction of the pilot with the systems of limiting the maximum flight modes. In: Scientific and methodological materials on ensuring flight safety. Moscow: Air Force Engineering Academy, named by N.E. Zhukovsky
12. Akimov AN, Andreev VV (1997) Method of direct optimization. In: The main provisions and procedure of application. – M. - Air Force Engineering Academy, named by N.E. Zhukovsky
13. Smernicki VF (1998) Scientific and methodological basis for the construction and operation of on-board integrated systems for safety of flights. In: Problems of safety flights. M.: VINITI, pp 25–32

Chapter 2
The Method of Adaptive Maintenance of Constraints on the Components of the State Vector of a Dynamic System

2.1 Substantial Problem Statement

Consider an arbitrary dynamic object, whose behavior is described by some system of n ordinary nonlinear differential equations in the Cauchy form [1]:

$$\dot{x} = f(x, a, u, t), \tag{2.1}$$

where x—n-dimensional state vector of a dynamical system, defined in space X^n; a—r-dimensional vector of parameters that takes values from the set A^r and are characterized by the properties of the object and the environment; u—m-dimensional vector of control actions, formed by the control system and belonging to the set U^m; t—current time belonging to the segment $[t_0, t_k]$, on which the equation is defined (2.1); $f(\cdot)$—the n-dimensional vector function of the specified arguments.

For the system (2.1), the initial conditions are set as

$$x(t_0) = x_0$$

and arbitrary control

$$u = u(t). \tag{2.2}$$

Let us assume that in space X^n there is a certain stationary constraint surface, which is convex with respect to the current state of the dynamical system

$$\xi(x) = 0, \tag{2.3}$$

dividing the space X^n into operational and non-operational areas. And is known, the «piercing» of the surface (2.3) by the trajectory of the movement of dynamical system $x(t)$ is unacceptable from the conditions of its existence.

V. Akimov and V. A. Zatuchny, *Limiting Modes of Aircraft Flight*,
Springer Aerospace Technology, https://doi.org/10.1007/978-981-19-6329-2_2

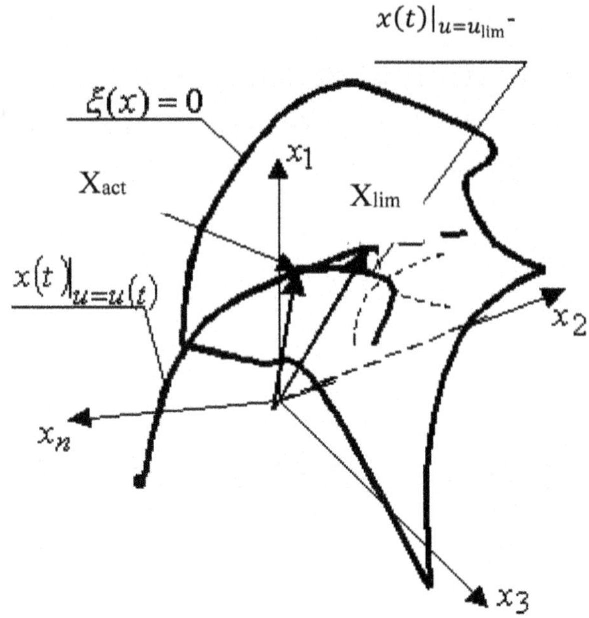

Fig. 2.1 Trajectories of motion of a dynamic system in space X^n

The task of automating the maintenance of the constraint, described by Eq. (2.3) in the state space X^n with the help of a pre-defined, relay-activated control $u_{\text{orp}} \in U^m$, is set. For this purpose, it is necessary to determine a moment in time or a point in the state space X^n, depending on the specified control of the constraint, the rate of approach to the surface, as well as the internal parameters of the dynamic system itself, in which the selected control $u_{\lim}$ ensures that the trajectory of the movement of the dynamic system (DS) touches the surface of the constraint without piercing the latter (see Fig. 2.1).

The following are marked in Fig. 2.1: $x(t)|_{u=u(t)}$, $x(t)|_{u=u_{\lim}}$—trajectories of movement DS in space X^n with arbitrary control and constraint control, respectively; x_{act}—a vector, characterized by a point on the trajectory of the dynamic system, in which the constraint system is activated; $x_{\lim}$—limit state vector DS on the trajectory $x(t)|_{u=u(t)}$.

2.2 General Case of Keeping the Constraint

Consider a pre-normalized space in some way X^n.

Let in this space be given: point $M_0(x)$ with the current coordinates DS, direction and speed of its movement $\vec{\dot{x}}$, and the constraint surface, described by the Eq. (2.3). The geometric interpretation of the problem for a two-dimensional space is presented in Fig. 2.2.

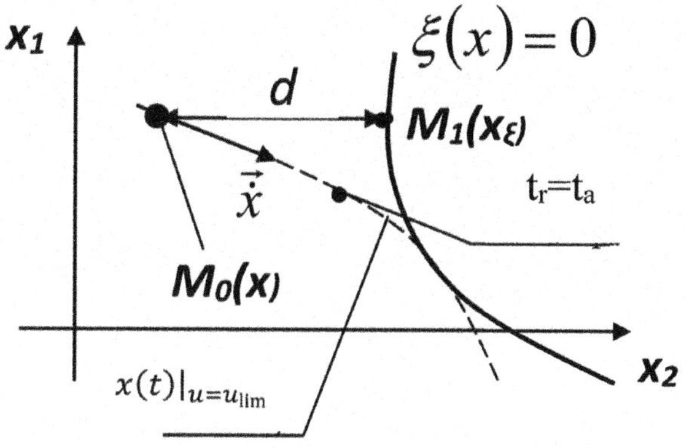

Fig. 2.2 Geometric interpretation of the problem for a two-dimensional space

We introduce the concept of range d (see Fig. 2.2), by which we mean the shortest distance between a point $M_0(x) \in X^n$ and constraint surface $\xi(x) = 0$.

In general, the range d can be determined by minimizing the criterion function

$$F(x) = d^2 = \left|x - x_\xi\right|^2, \tag{2.4}$$

where x_ξ—vector, whose coordinates characterize the position of an arbitrary point $M_1(x_\xi)$ lying on the surface $\xi(x) = 0$. The coordinates of the vector x_ξ can also be determined, using the equation of the normal to the constraint surface passing through the point $M_0(x)$, based on the obvious fact that the line containing the shortest distance from the point $M_0(x)$ to the constraint surface coincides with the normal to the constraint surface at the point $M_1(x_\xi)$.

It should be noted that any displacement in the space X^n of the point $M_0(x)$ causes a corresponding displacement over the constraint surface of the point $M_1(x_\xi)$ satisfying minF(x). Therefore, the following formal relationship takes place:

$$d = d\Big(x, x_\xi, (x)\big|_{\min F(x)}\Big). \tag{2.5}$$

In the future, the index min$F(x)$ will be omitted.

Let us consider, in terms of distance, the procedure for reaching the boundary $\xi(x) = 0$, taking into account that its not piercing. To do this, we will analyze two characteristic time intervals. The first interval is the time to reach the DS of the constraint surface. Let's denote it as t_r. The second interval is the available time t_a, during which the rate of approach to the constraint surface can be extinguished to zero. Both time intervals are defined under the assumption that constraint control is implemented for the DS:

$$u = u_{\text{lim}}. \tag{2.6}$$

It is natural to assume that when equality is achieved

$$t_r = t_a, \tag{2.7}$$

the trajectory of the DS with the control (2.6), turned on at time (2.7), will only touch the surface of the constraint (see Fig. 2.2).

We express the values t_r and t_a in terms of the range d and its derivatives under the assumption that the function (2.4) is defined and differentiable over the time interval under consideration. Then the time to reach the DS surface of the constraint will be equal to the ratio:

$$t_{\partial} = -\frac{d}{\dot{d}_{\text{mean}}}, \tag{2.8}$$

where $\dot{d}_{\text{mean}} < 0$—mean speed of approach of a point $M_0(x)$ with a constraint surface $\xi(x) = 0$.

Given the condition that

$$\dot{d}\big|_{d=0} = 0 \tag{2.9}$$

and making an assumption about the linearity of the speed reset from the current value $\dot{d}$ to zero, by providing $\ddot{d} = const$, it can be shown that

$$\dot{d}_{\text{mean}} = \frac{\dot{d} + 0}{2} = \frac{\dot{d}}{2}. \tag{2.10}$$

Substituting (2.10) in (2.8), for the time of reaching t_r, we have the following dependence:

$$t_r = -\frac{2d}{\dot{d}}. \tag{2.11}$$

By analogy with the expression (2.8), in the space of higher order derivatives, the available time t_a, in accordance with its physical meaning, is determined by the expression [2]:

$$t_a = -\frac{\dot{d}}{\ddot{d}_a}. \tag{2.12}$$

Here, $\ddot{d}_a$—available deceleration acceleration of the dynamic system under the action of the constraint control (2.6).

It should be noted that for the correct solution to the problem of maintaining the constraints in this statement, as well as for the time of achievement, it is necessary to provide

$$\ddot{d}_a = \ddot{d}\big|_{u=u_{\text{lim}}} = \text{const}. \tag{2.13}$$

Equating Eqs. (2.11) and (2.12), and solving them with respect to the range d, we obtain a dependence for determining the distance to the surface of the constraint $d_{\lim}$, at which it is necessary to turn on the control (2.6). Having done the corresponding transformations, we have

$$d_{\lim} = \dot{d}^2/2\ddot{d}_a. \tag{2.14}$$

In (2.14), $\dot{d}$ and $\ddot{d}_a$ are determined by formally finding the first and second derivatives (taking into account (2.6)) of the dependence (2.5) on t as a parameter. Having done this procedure, we have

$$\dot{d} = \left(\frac{\partial d}{\partial x^m} + \frac{\partial d}{\partial x_\xi^m}\frac{\partial x_\xi}{\partial x^m}\right) f(x, a, u, t) \tag{2.15}$$

and in a linear setting:

$$\ddot{d}_a = \left(\frac{\partial d}{\partial x^m} + \frac{\partial d}{\partial x_\xi^m}\frac{\partial x_\xi}{\partial x^m}\right)\left(\frac{\partial f(\cdot)}{\partial t} + \frac{\partial f(\cdot)}{\partial x^?} f(x, a, u_{\lim}, t)\right) \tag{2.16}$$

(summand $\left\{\frac{d}{dt}\left(\frac{\partial d}{\partial x^m} + \frac{\partial d}{\partial x_\xi^m}\frac{\partial x_\xi}{\partial x^m}\right)\right\} f(x, a, u, t)$ for simplifying further calculations in (2.16) does not take into account).

The main problem with calculating the values of $\dot{d}$ and $\ddot{d}_a$ by (2.15), (2.16) is to obtain an analytical dependence for partial derivatives $\frac{\partial d}{\partial x^m}, \frac{\partial d}{\partial x_\xi^m}, \frac{\partial x_\xi}{\partial x^m}$, the physical meaning of which is illustrated in Fig. 2.3 (in the figure, the distance increment $\Delta d_{\Delta x}$ caused by moving a point $M(x)$ on value Δx, and increment $\Delta d_{\Delta x_\xi}$—offset on the surface $\xi(x) = 0$ of point $M(x_\xi)$ by an amount Δx_ξ). This problem lies in the fact that in the procedure of their calculation, there is a constantly performed operation of minimizing the function (2.4). If the dynamic system moves in space X^n strictly normal to the constraint surface $\xi(x) = 0$, then $x_\xi = const$ and the first factor in (2.15) and (2.16) is equal to $\left(\frac{\partial d}{\partial x^m} + \frac{\partial d}{\partial x_\xi^m}\frac{\partial x_\xi}{\partial x^m}\right) = \frac{(x - x_\xi)}{|x - x_\xi|}$.

The way to solve this problem will be presented in the following paragraphs.

Analysis of the expression (2.14) and its components (2.15), (2.16) shows the following.

First, when determining the $d_{\lim}$ section, the following factors are taken into account:

- custom properties DS—$\frac{\partial f(\cdot)}{\partial x^m}$,
- its non-stationarity—$\frac{\partial f(\cdot)}{\partial t}$,
- environment parameters—a,
- the component of the change in the distance to the constraint surface, taking into account the curvature of the latter, in the case of moving DS along an arbitrary spatial line that does not coincide with the normal—$\frac{\partial d}{\partial x\xi^m}\frac{\partial x\xi}{\partial x^m}$,
- trusted control restrictions—$u_{\lim}$.

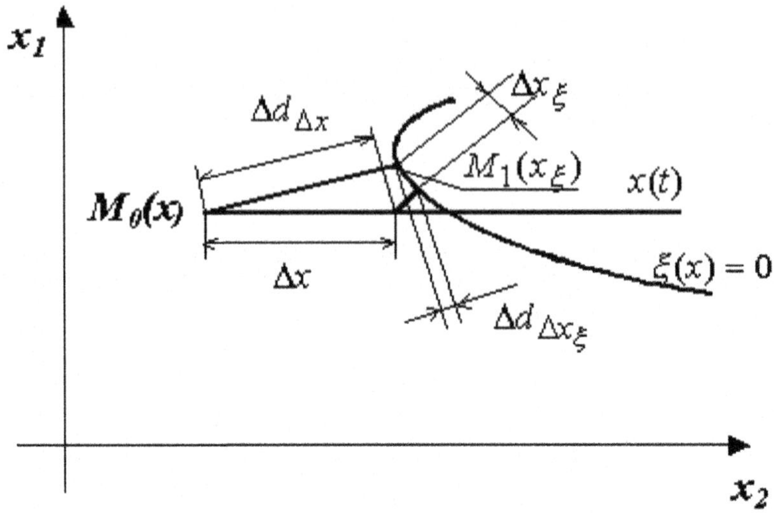

Fig. 2.3 The physical meaning of partial derivatives

All this allows us to conclude that there are adaptive properties of the algorithm for maintaining constraints to the configuration of the constraint surface, and the properties of the medium itself, DS.

Second, from (2.16) we can assume that under the action of the control constraints (2.6) will change $\ddot{d} = \ddot{d}_a$ in the restricted area, which is contrary to the fundamental condition (2.13). For his unconditional implementation, it is necessary to introduce compensation control $\Delta u(t)$ in (2.6), that is,

by putting

$$u = u_{\lim} + \Delta u(t),$$

to provide

$$\Delta\ddot{d}(t) = \ddot{d}_a(t) - \ddot{d}_{a_{mean}} = \left(\frac{\partial d}{\partial x^{\mathrm{T}}} + \frac{\partial d}{\partial x_\xi^{\mathrm{T}}}\frac{\partial x_\xi}{\partial x^{\mathrm{T}}}\right)\frac{\partial f(\cdot)}{\partial x^T}\left[f(\cdot) - f_{cp}(\cdot)\right] \equiv 0, \quad (2.17)$$

where $\ddot{d}_{a_{\text{mean}}}$, $f_{\text{mean}}(\cdot)$—the available deceleration acceleration and the vector function $f(\cdot)$ at the moment, when the restriction system is activated, respectively.

In this case, (2.16) can be rewritten as

$$\Delta\ddot{d}(t) = \frac{\partial\ddot{d}}{\partial u^{\mathrm{T}}}\Delta u(t) = \left(\frac{\partial d}{\partial x^{\mathrm{T}}} + \frac{\partial d}{\partial x_\xi^{T}}\frac{\partial x_\xi}{\partial x^{\mathrm{T}}}\right)\frac{\partial f(\cdot)}{\partial x^T}\frac{\partial f(\cdot)}{\partial u^{\mathrm{T}}}\Delta u(t). \quad (2.18)$$

From there, by expressing $\Delta u(t)$:taking into account (2.17), we obtain an expression for the compensation control:

$$\Delta u(t) = \left[\frac{\partial f(\cdot)}{\partial u^m}\right]^{\#} [f(\cdot) - f_{\text{mean}}(\cdot)]. \tag{2.19}$$

Here and further, the sign # corresponds to the operation of pseudo-inversion of non-square and degenerate matrices [3]. It should be noted that the value of the compensation control $\Delta u(t)$ does not depend on the parameters of the constraint surface and the distance to it, but depends only on the properties of the dynamic system itself.

As a result, for the adaptive system of maintaining the restriction, we have the following algorithm for its activation:

$$u = \begin{cases} u(t), \text{ если } d > d_{\text{lim}}; \\ u_{\text{lim}} + \Delta u(t), \text{ если } d \le d_{\text{lim}}. \end{cases} \tag{2.20}$$

In (2.20), u_{lim} has the meaning of a pre-selected control action, trusted to the system of maintaining the constraints, and $\Delta u(t)$ provides the current adjustment of the algorithm to the non-calculated conditions of the DS operation.

2.3 Proof of the Analogy of the Method of Maintaining Constraints to the Fundamental Law of Equal-Variable Motion

Consider the problem of maintaining constraints for an object, represented by a system of second-order differential equations:

$$\begin{cases} \dot{X} = V; \\ \dot{V} = a. \end{cases} \tag{2.21}$$

The model (2.21) describes the motion of a material point in a given direction. Here, X is the change in the linear coordinate of the object, V is the velocity, and a is the acceleration.

The system (2.21) is solved with the initial conditions:

$$X(0) = X_0, \quad V(0) = V_0.$$

The problem of maintaining the constraints is formulated as follows. You need to turn on control

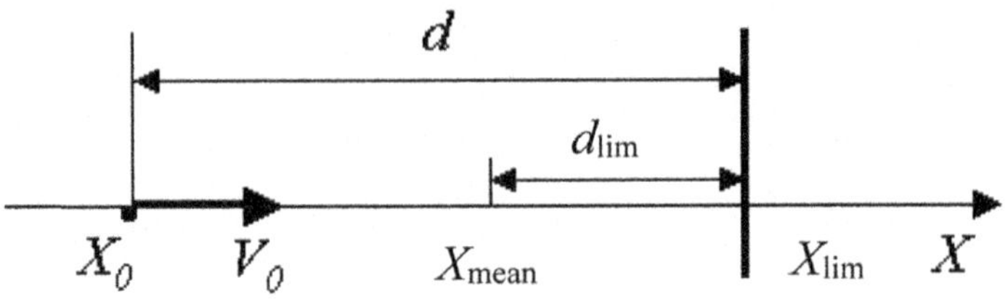

Fig. 2.4 The moment, when the object reaches a certain coordinate $X = X_{\text{mean}}$, such that the limit is not exceeded, $X = X_{\text{lim}}$

$$a = a_{\text{lim}} = const \tag{2.22}$$

at the moment, when the object reaches a certain coordinate $X = X_{\text{mean}}$, such that the limit is not exceeded, $X = X_{\text{lim}}$ (see Fig. 2.4).

The same problem can be formulated differently. For an arbitrary speed V, find the coordinate of the point $X = X_{\text{mean}}$, at which the speed under the action of the specified control $a = a_{\text{lim}} = const$ will be reset to zero on the level $X = X_{\text{lim}}$.

Taking into account the known general solution of the system (2.21)

$$X(t) = X_0 + V_0 t + \frac{at^2}{2},$$

$$V(t) = V_0 + at, \tag{2.23}$$

for the braking section $X(t) > X_{\text{mean}}$ define the value of the coordinate $X = X_{\text{lim}}$

$$X_{\text{lim}} = X_{\text{mean}} + V\Delta t + \frac{a_{\text{lim}}\Delta t^2}{2}, \tag{2.24}$$

where Δt—the time, when the object passes the interval $[X_{\text{mean}}, X_{\text{lim}}]$; V—the current speed of the object. As a result, we have two unknowns X_{mean} and Δt. Since by the condition

$$V|_{X=X_{\text{lim}}} = 0, \tag{2.25}$$

we define (2.24) by the equation:

$$V + a_{\text{lim}}\Delta t = 0. \tag{2.26}$$

Equality (2.26) is obtained as a result of solving the second equation of the system (2.21) and is written taking into account (2.25) for the interval $[X_{\text{mean}}, X_{\text{lim}}]$. By expressing from (2.26) Δt and substituting it into (2.24), we can obtain an expression for the control turning on point of the constraint $a = a_{\text{lim}}$:

$$X_{\text{mean}} = X_{\text{lim}} + \frac{V^2}{2a_{\text{lim}}}. \tag{2.27}$$

Denoting $d = X_{\text{lim}} - X$; $d_{\text{lim}} = X_{\text{lim}} - X_{\text{mean}}$; $\dot{d} = -V$; $\ddot{d}_a = -a_{\text{lim}}$, we rewrite (2.27) in terms of the range. As a result,

$$d_{\text{lim}} = \dot{d}^2/2\ddot{d}_a \tag{2.28}$$

It is easy to see that the expression (2.28) completely coincides with (2.14).

Thus, in the example of a double integrating link, an exact analytical solution for calculating the length of the restriction section is obtained, which proves the validity of the calculations, given for the general case.

Results of modeling the process of maintaining the restriction $X = X_{\text{lim}}$ for the model (2.21) are presented in Fig. 2.5 (the case of uniform motion) and Fig. 2.6 (uniformly accelerated motion), and confirm the efficiency of the method. In both cases, when the object reaches the value $X = X_{\text{lim}}$, the control was assigned the value $a = 0$.

In the future, the control task will be limited to the process of preventing going beyond the controlled parameter or parameters. The problem of stabilization at the level of the limited value of the parameter will not be considered.

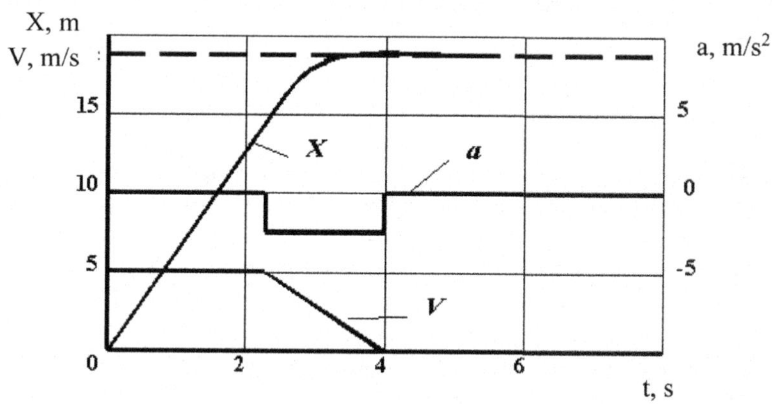

Fig. 2.5 Results of modeling the process of maintaining the restriction for uniform movement

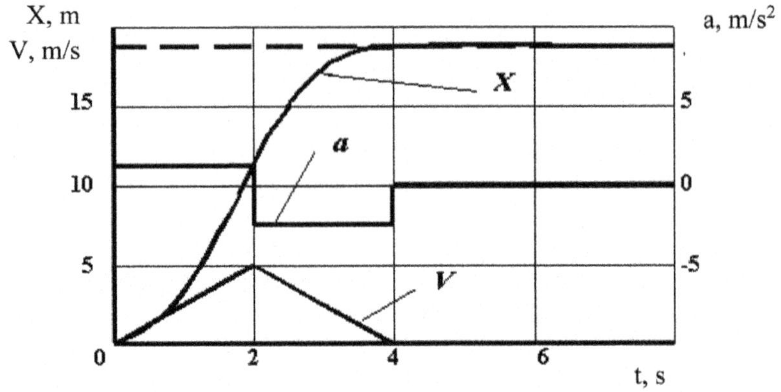

Fig. 2.6 Results of modeling the process of maintaining the restriction for uniformly accelerated motion

2.4 Synthesis of Algorithmic Support for an Adaptive Constraint System for Particular Models of a Dynamic System

2.4.1 *Some Ways to Describe Stationary Constraint Surfaces*

It was shown above that the proposed method of maintaining constraints is based on analysis in metric space X^n such an integral parameter as the distance d between the point $M_0(x)$, which characterizes the current state of the dynamic system, and the constraint surface $\xi(x) = 0$. The greatest difficulty in using the described method is obtaining an analytical functional dependence for the parameter d (2.5) and finding its derivatives (2.15), (2.16), when considering complex constraint surfaces. To solve this problem, the following approaches can be proposed:

(1) Constraint surface $\xi(x) = 0$ is approximated by a polyhedron, whose faces are hyperplanes, described by a matrix equation of the form:

$$a_s x + c_s = 0, \tag{2.29}$$

where a_s, c_s—the matrix is a string and a scalar, respectively, characterizing the position of the s-th hyperplane in space; s—the number of the face of the polyhedron. An illustration of this case for a two-dimensional space is shown in Fig. 2.7.

Then, in accordance with the analytical geometry in space [4], the distance to the hyperplane through which the piercing is predicted (in Fig. 2.7, this face—s) can be defined using an elementary expression:

$$d_s = \overline{a}_s x + \overline{c}_s. \tag{2.30}$$

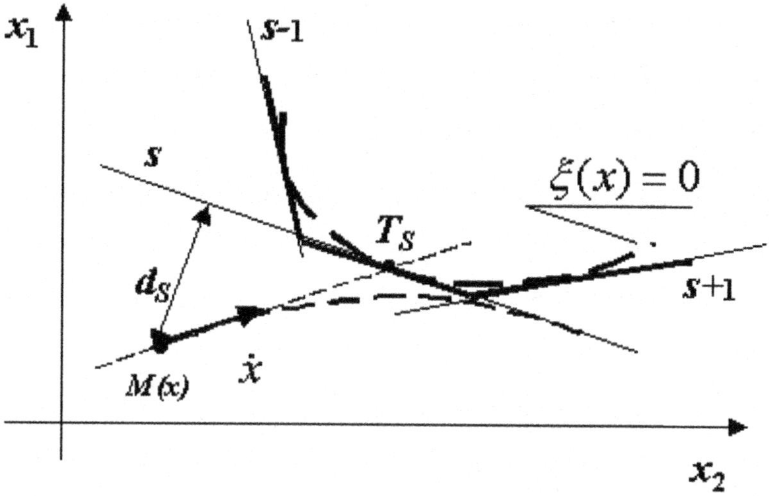

Fig. 2.7 Illustration of a constraint surface for a two-dimensional space

In (2.30), $\overline{a}_s = a_s/|a_s|$; $\overline{c}_s = c_s/|a_s|$; $|a_l|$—the module of the vector.

The simplicity of this approach is obvious. However, its implementation is problematic due to the need to analyze an innumerable set of hyperplanes and the intervals of their existence. Reducing the number of approximating hyperplanes will reduce the accuracy of maintaining the boundary $\xi(x) = 0$ (see Fig. 2.8).

(2). The other approach [5], from the authors' point of view, is more preferable. It is based on the current analysis of the DS movement's prehistory. Using this analysis,

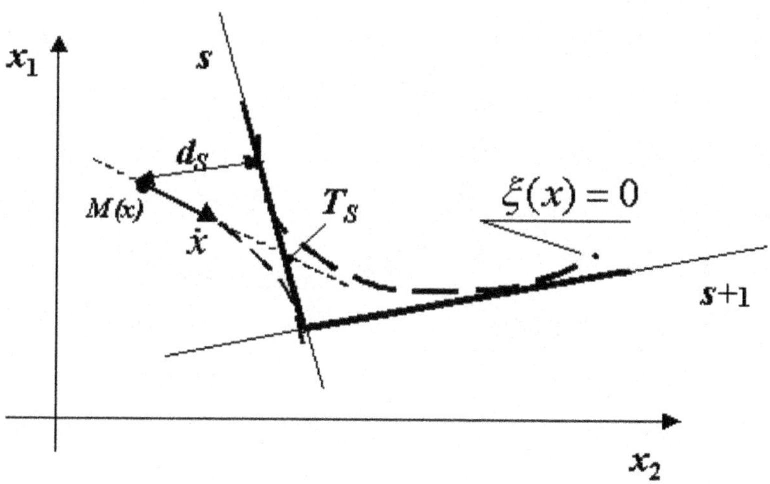

Fig. 2.8 Illustration of the relationship between the number of approximating hyperplanes and the accuracy of maintaining the boundary $\xi(x) = 0$

you can set the equation of trajectory, for example, in the form

$$g(x) = 0 \tag{2.31}$$

(in the simplest case, it can be the equation of the spatial line).

Solving together the system of equations of the trajectory and the constraint surface

$$\begin{cases} g(x) = 0; \\ \xi(x) = 0 \end{cases} \tag{2.32}$$

we can define the coordinates x_m of point T_1 predicted penetration of the constraint surface by the trajectory of the DS movement (see Fig. 2.9).

Let's build at the point T_1 the tangent hyperplane to the constraint surface. Its equation will have the form:

$$\partial\xi(x_m)/\partial x^m (x - x_m) = 0. \tag{2.33}$$

Let's introduce the notation:

$$a = \frac{\partial\xi(x_m)}{\partial x^m}; \quad c = -\frac{\partial\xi(x_m)}{\partial x^m} x_m,$$

where a—matrix-string, c—scalar, and we obtain the equation of the tangent of the hyperplane to the constraint surface in the general form:

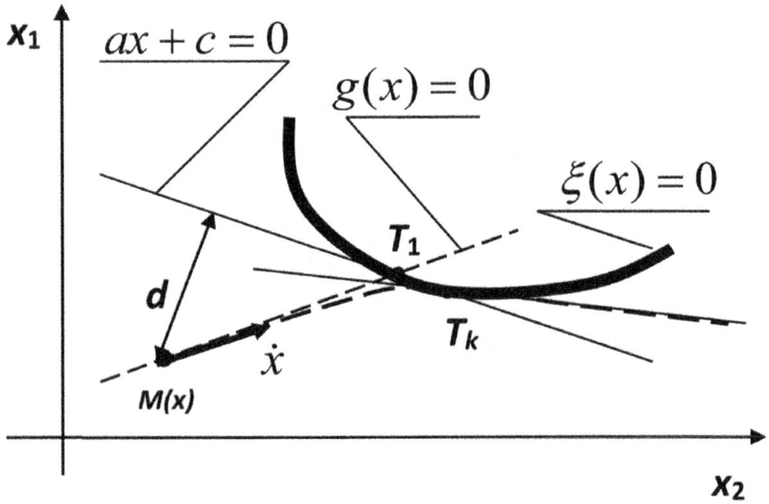

Fig. 2.9 Method for determining the coordinate of the point of the predicted piercing by the trajectory of the dynamic system's movement of the constraint surface

$$ax + c = 0. \tag{2.34}$$

In the future, we will use the expression (2.34) as the equation of the hyperplane of the constraint. It should be noted that, when the trajectory of movement changes in the process of evading the constraint surface, the coordinates of the point T_i, belonging to the surface $\xi(x) = 0$, will also change, therefore, the position in the space of the boundary hyperplane will also change.

Thus, the cyclic recalculation of the hyperplane parameters will allow the constraint algorithms to provide the properties of adaptivity to the shape of the constraint surface.

In Fig. 2.9, the point T_1 characterizes the initial position of the tangent to the surface $\xi(x) = 0$ and the point T_k characterizes its final position.

As a result, the current range to the hyperplane (2.34) is defined as

$$d = \overline{a}x + \overline{c}, \tag{2.35}$$

where $\overline{a} = a/|a|;\quad \overline{c} = c/|a|$.

2.4.2 *Algorithm for Constraining a Deterministic Linear Continuous Dynamic System*

We consider a system, whose state change in time is described by the equation:

$$\dot{x} = A(t)x + B(t)u. \tag{2.36}$$

In (2.36), $x \in X^n$; $u \in U^m$; $A(t)$, $B(t)$—coefficient matrices with the size of $(n \times n)$ и $(n \times m)$, respectively, in the general case are functions of time. If $A = const$, $B = const$, then system (2.36) is stationary.

We synthesize for the model (2.36) the algorithm of the constraint system, described by expressions (2.20), taking into account (2.14). For this purpose, we derive the formula dependencies of its parameters. Under the parameters of the restriction system, we will understand the components, included in the expressions (2.14) and (2.20), that is, the first and second derivatives of the range d, as well as the value of the compensation control of the restriction $\Delta u(t)$.

We find $\dot{d}$ and $\ddot{d}$ by sequentially differentiating by t the expression (2.36):

$$\dot{d} = \overline{a}\dot{x} = \overline{a}[A(t)x + B(t)u]; \tag{2.37}$$

$$\ddot{d} = \overline{a}\frac{d}{dt}[A(t)x + B(t)u] = \overline{a}\left\{\begin{array}{l}\left[\dfrac{dA(t)}{dt} + A(t)A(t)\right]x \\ +\left[\dfrac{dB(t)}{dt} + A(t)B(t)\right]u + B(t)\dot{u}\end{array}\right\}. \tag{2.38}$$

Putting in (2.38) $u = u_{\lim}$, we determine the available deceleration acceleration DS for its evasion from the constraint surface, given by the hyperplane (2.35). As a result,

$$\ddot{d}_a = \overline{a}\left\{\left[\frac{dA(t)}{dt} + A(t)A(t)\right]x + \left[\frac{dB(t)}{dt} + A(t)B(t)\right]u_{\lim}\right\}. \tag{2.39}$$

Note that (2.39) takes into account: the configuration of the original constraint surface through the matrix $\bar{a}$; non-stationarity DS—matrix $\frac{dA(t)}{dt}$ and $\frac{dB(t)}{dt}$; custom properties DS—matrix $A(t)$; efficiency of control—matrix $B(t)$. Changing the listed matrices for some reason will only lead to a change in the point of turning on the constraint system in the operation, defined by the expression (2.14). Thus, here we can also talk about the adaptability of the constraint algorithms to the parameters of the constraint surface and the properties of DS.

Fulfilling the condition $\ddot{d} = const$ at the stage of limitations ($d < d_{\lim}$), it is provided by compensatory control, obtained by analogy with the output of the Eq. (2.19):

$$\Delta u(t) = \left[\frac{dB(t)}{dt} + A(t)B(t)\right]^{\#}\left[\frac{dA(t)}{dt} + A(t)A(t)\right](x(t) - x_{mean}). \tag{2.40}$$

For the case, if the system (2.36) is stationary, the expressions (2.37), (2.39), and (2.40) will take a simpler form:

$$\dot{d} = \overline{a}(Ax + Bu); \tag{2.41}$$

$$\ddot{d}_= \overline{a}A(Ax + Bu_{\lim}); \tag{2.42}$$

$$\Delta u(t) = B^{\#}A(x(t) - x_{mean}). \tag{2.43}$$

The resulting expressions can be used in the development of algorithms for limiting flight parameters AC [6, 7].

2.4.3 An Algorithm for Constraining a Deterministic Continuous Controlled Multilevel Dynamical System

A wide class of dynamical systems, including the description of the trajectory motion of AC, allows the description to use a multilevel system of vector differential equations in the form of Cauchy. Consider a two-level DS, represented as

$$\begin{cases} \dot{x} = f(x, z, t); \\ \dot{z} = \phi(z, u, t), \end{cases} \tag{2.44}$$

where $x \in X^n$—the state vector DS; $z \in Z^l$—pseudo-control vector for the vector x; $u \in U^m$—vector of valid controls; $f(\cdot)$ и $\phi(\cdot)$—vector functions of vector and scalar arguments of the corresponding dimension. The system (2.44) can be obtained by decomposing the model (2.1) by the method, described in [5].

Let us consider obtaining the main computational dependencies for an adaptive constraint system, the algorithm of which is represented by the condition (2.20).

Let the boundary hyperplane, which is tangent to the constraint surface $\xi(x) = 0$, defined in space X^n, be described by Eq. (2.34) and the distance to it be described by Eq. (2.35). Find the first and second derivatives of the distance d, value $\ddot{d}_a$, and compensation control $\Delta u(t)$ also. Differentiating the expression (2.35) by t and substituting the right part of the first equation of the system (2.44) instead of x, we have

$$\dot{d} = \bar{a} f(\cdot). \tag{2.45}$$

Re-differentiating with respect to t, taking into account (2.44), gives the following expression:

$$\ddot{d}(t) = \bar{a}\left[\frac{\partial f(\cdot)}{\partial t} + \frac{\partial f(\cdot)}{\partial x^m} f(\cdot) + \frac{\partial f(\cdot)}{\partial z^m}\phi(\cdot)\right]. \tag{2.46}$$

Putting in a vector function $\phi(\cdot)$ of Eq. (2.46) $u = u_{\text{lim}}$, we get

$$\ddot{d}(t) = \bar{a}\left[\frac{\partial f(\cdot)}{\partial t} + \frac{\partial f(\cdot)}{\partial x^m} f(\cdot) + \frac{\partial f(\cdot)}{\partial z^m}\phi(z, u_{\text{lim}}, t)\right]. \tag{2.47}$$

And, finally, writing in increments the expression (2.46) and expressing the value $\Delta u(t)$ from it, we have

$$\Delta u(t) = \left[\bar{a}\frac{\partial f(\cdot)}{\partial z^m}\frac{\partial \phi(\cdot)}{\partial u^m}\right]^{\#} \Delta\ddot{d}(t) \tag{2.48}$$

where

$$\Delta\ddot{d}(t) = \ddot{d}_a(t) - \ddot{d}_{a_{\text{mean}}}.$$

As a result, for the algorithm of the system of maintaining restrictions (2.20), its components are obtained, namely $\dot{d}$, $\ddot{d}$, $\Delta u(t)$.

Thus, in the spectrum of the description of dynamic systems by deterministic continuous models, an algorithmic basis for solving the problems of adaptive restriction of the state vector of the DS is obtained. The following chapters are devoted to specific cases of limiting the movement parameters and the totality of such a complex dynamic system as an aircraft.

References

1. Byushgens GS, Studnev RV (1979) Aerodynamics of aeroplane. Dynamics of longitudinal and lateral motion. M, Mashinostroenie
2. Wentzel ES, Ovcharov LA (1988) Probability theory and its engineering applications. Moscow, Nauka
3. Vygodsky MY (1958) Handbook of higher mathematics. M, State Publishing house of technical and theoretical literature
4. Vorobyov V. G., Zubkov B. V., Urinovsky B. D. Technical means and methods of ensuring flight safety. - M.: Transport, 1989.
5. Akimov A.N., Andreev V.V. Direct optimization method. The main provisions and the order of application. - M: Air Force Engineering Academy named after N. E. Zhukovsky, 1997.
6. Zatuchny DA (2007) Construction of the optimal trajectory of the AC movement according to the proposed criterion. Scientific Bulletin of MSTU GA, series «Navigation and ATC», No 121, , pp 156–158
7. Zatuchny DA, Logvin AI (2013) Criterion for reducing the dynamic error in the automatic dependent observation mode. Scientific Bulletin of MSTU GA, No 189, pp 5–8

Chapter 3
Algorithms for Adaptive Limitation of Aircraft Flight Parameters

3.1 Adaptive Limitation of Aircraft Angles of Attack

In the first section, the disadvantages of the system of limiting the angle of attack with a fixed parameter setting were considered. We apply the method of adaptive restriction of the components of the state vector of a dynamic system to the problem of limiting the angle of attack in an isolated longitudinal motion.

Let the longitudinal short-period motion of the aircraft be described by a system of second-order equations of the form:

$$\dot{\alpha} = \omega_z - \overline{Y}^{\alpha}\alpha;$$
$$\dot{\omega}_z = \overline{M}_z^{\alpha}\alpha + \overline{M}_z^{\dot{\alpha}}\dot{\alpha} + \overline{M}_z^{\omega_z}\omega_z + \overline{M}_z^{\delta_e}\delta_e \tag{3.1}$$

In (3.1), all notations are generally accepted.

To work out the methodological issues of the synthesis of algorithms for the restriction system, the aircraft control system is set by a set of controls from the pilot, the pitch damper, and the automatic longitudinal stability in the angle of attack:

$$\delta_e = K_e x_e + K_e^{\omega_z}\omega_z + K_e^{\alpha}\alpha. \tag{3.2}$$

The coefficients in (3.2) are functions of the flight mode, i.e.

$$K_e = K_e(H, M),\ K_e^{\omega_z} = K_e^{\omega_z}(H, M),\ K_e^{\alpha} = K_e^{\alpha}(H, M)$$

.

We assume that the constraint $\alpha = \alpha_{\lim}$ in the coordinates (α, ω_z) has the form, shown in Fig. 3.1. It can also be described by the matrix equation:

$$ax + c = 0, \tag{3.3}$$

V. Akimov and V. A. Zatuchny, *Limiting Modes of Aircraft Flight*,
Springer Aerospace Technology, https://doi.org/10.1007/978-981-19-6329-2_3

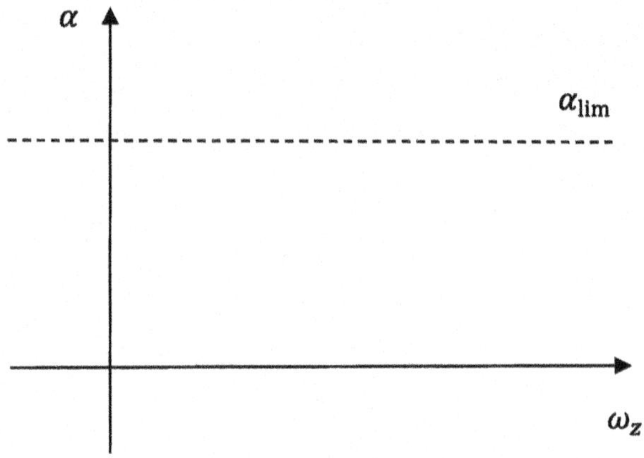

Fig. 3.1 Type of restriction $\alpha = \alpha_{\lim}$ in coordinates (α, ω_z)

where $x^m = \left[\alpha\ \omega_z\right]$; $a = [-1\quad 0]$; $c = \alpha_{\lim}$.

In this case, the excess of $\alpha_{\lim}$ is unacceptable from the conditions for ensuring the safety of the flight of the aircraft [1, 2].

The distance from the point that characterizes the current state of the constraint object to the boundary (3.3) is determined by the expression (2.35). Making the appropriate substitutions, for the case under consideration, we obtain the obvious equality:

$$d = \alpha_{\lim} - \alpha. \tag{3.4}$$

We will set the constraint control in the form, corresponding to the law (2.20):

$$x = [\alpha\ \beta\ \omega_x\ \omega_y\ \omega_z]^m;\ u = [\delta_e\ \delta_n]^m. \tag{3.5}$$

Here $x_{e_{\lim}}$—the deviation of the aircraft control stick trusted to the restriction system; $\Delta x_e(t)$—compensation control; $d_{\lim}$—distance to the border $\alpha_{\lim}$; when it is necessary to turn on the constraint control, it is determined by the dependency (2.14).

In accordance with the above method, we find the first and second derivatives of the distance d, value $\ddot{d}_a$, as well as compensation control $\Delta x_e(t)$. It is obvious from (3.4) that

$$\begin{aligned} \dot{d} &= -\dot{\alpha}; \\ \ddot{d} &= -\ddot{\alpha}. \end{aligned} \tag{3.6}$$

To obtain the component of the Eq. (2.14) $\ddot{d}_a = -\ddot{\alpha}\left(x_{e_{\lim}}\right)$, we will proceed as follows. Let's rewrite the system of Eqs. (3.1) in the form of «input–output». Then taking into account (3.2) and substituting for x_e control of restrictions $x_{e_{\lim}}$, we have

$$\ddot{d}_a = -\left(\overline{M}_z^{\dot{\alpha}} - \overline{Y}^{\alpha} + \overline{M}_z^{\omega_z} + \overline{M}_z^{\delta_e} K_e^{\omega_z}\right)\dot{\alpha}$$

$$- \left(\overline{M}_z^{\alpha} + \left(\overline{M}_z^{\omega_z} + \overline{M}_z^{\delta_e} K_e^{\omega_z}\right)\overline{Y}^{\alpha} + \overline{M}_z^{\delta_e} K_e^{\alpha}\right)\alpha - \overline{M}_z^{\delta_e} K_e x_{e_{\lim}}. \tag{3.7}$$

As a result, the expression that defines the moment, when the system of the constraint of the angle of attack is activated, will take the following form:

$$d_{\lim} = K\dot{\alpha}^2, \tag{3.8}$$

where coefficient $K = var$

$$K = \frac{1}{2}\frac{1}{-\left(\overline{M}_z^{\dot{\alpha}} - \overline{Y}^{\alpha} + \overline{M}_z^{\omega z} + \overline{M}_z^{\delta e} K_e^{\omega z}\right)\dot{\alpha} - \left(\overline{M}_z^{\alpha} + \left(\overline{M}_e^{\omega z} + \overline{M}_z^{\delta e} K_e^{\omega z} \overline{Y}^{\alpha}\right) + \overline{M}_z^{\delta e} K_e^{\alpha} \alpha\right) - \overline{M}_z^{\delta e} K_e x_{e\text{orp}}}$$

and depends on the following:

- own properties of the object itself, taking into account the system for improving stability and controllability;
- trusted control restriction to the system;
- the efficiency of control;
- current values α and $\dot{\alpha}$.

Note that the structure of the expression (3.8) differs significantly from the traditionally accepted one [3, 4]

$$d_{\lim} = K\dot{\alpha}, \tag{3.9}$$

where $K = const$.

First, instead of the first power of the rate of change of the angle of attack in (3.9), the square of this value is used in (3.8). Second, the K coefficient became a calculated coefficient. Moreover, even if the natural properties of the constraint object, its stability, and controllability characteristics change as a result of possible failures, the constraint algorithm adapts accordingly to the new conditions. This is achieved by using the procedure for the current identification of failures, for example, using the method, described in detail in [5, 6].

In addition, to maintain $\ddot{d} = const$ at the restriction stage, the law of control (3.5) contains the summand $\Delta x_e(t)$, defining compensation control and ensuring $\Delta\ddot{d}(t) \equiv 0$ at the stage of restriction. Compensation control $\Delta x_e(t)$ can be obtained by rewriting (3.7) in increments, namely.

$$\Delta\ddot{d}(t) = \frac{\partial\ddot{d}}{\partial x_e}\Delta x_e(t).$$

After performing the corresponding transformations, we have

$$\Delta x_e(t) = -\frac{\Delta\ddot{d}(t)}{\overline{M}_z^{\delta_e} K_e} = -\frac{\left(\ddot{d}_e(t) - \ddot{d}_{e_{\text{mean}}}\right)}{\overline{M}_z^{\delta_e} K_e}. \tag{3.10}$$

It should be noted that similar expanded expressions for $\dot{d}$, $\ddot{d}_a$, and $\Delta x_e(t)$ you can get it by using dependencies (2.41), (2.42), and (2.43).

The use of (3.5), (3.8), and (3.10) provides a strict solution to the problem of limiting the angle of attack with full consideration of all the main factors, affecting the dynamics of the movement of the AC.

Let us consider the model problem of limiting the angle of attack of the AC, using the above algorithms.

As a mathematical model of the control object, we will use the system (3.1), which operates together with the law of control (3.2).

Let the AC get control x_e, aimed at increasing the angle of attack, and there is a threat of going beyond α_{lim}. At each moment of time, in accordance with (3.4) and (3.8), the values d and d_{lim} are estimated and compared. If $d \leq d_{\text{lim}}$, the pilot is forcibly switched off from the circuit of control, and the control of restriction is turned on. Its implementation is described by Eqs. (3.5) and (3.10).

Turning off the constraint algorithm is possible, for example, when the condition $\dot{d} \geq 0$ is reached. In this case, the control of the aircraft is transferred to the balancing position $x_{e_{\text{bal}}}$. The value $x_{e_{\text{bal}}}$ can be obtained from (3.7) by equating it to zero and putting $\dot{\alpha} = 0, \ \alpha = \alpha_{\text{lim}}$. As a result,

$$x_{e_{\text{bal}}} = -\frac{\overline{M}_z^{\alpha} + \overline{M}_z^{\delta_e} K_e^{\alpha} + \left(\overline{M}_z^{\omega_z} + \overline{M}_z^{\delta_e} K_e^{\omega_z}\right)\overline{Y}^{\alpha}}{\overline{M}_z^{\delta_e} K_e} \alpha_{\text{lim}}. \tag{3.11}$$

Other conditions and techniques for turning off the system of the angle of attack limitation may also apply.

In Fig. 3.2, the results of solving the model problem of limiting the angle of attack at the level of 24° with different initial settings of the restriction system and ideal steering drives are presented.

Moreover, the graph displayed not the deviation of the control stick of the aircraft, but the deviation in time of the stabilizer δ_e. This was done to illustrate the operation of the stability and handling improvement system, along with the restriction system.

The figure shows that the lower the value of the restriction control,the angular position to which the stabilizer is transferred, the earlier the restriction system comes into operation and the longer the restriction process takes.

In Fig. 3.3, the results of modeling the process of limiting the angle of attack at the level of 24° at different initial flight speeds are presented. The adaptive properties of the algorithm to change the flight mode are obvious.

The algorithm for limiting robustness to the initial information about the own properties of the object was evaluated by coarsening its aerodynamic characteristics.

So, in Fig. 3.4, the results of modeling the process of limiting the angle of attack for different values of the aerodynamic coefficient m_z^{α} are presented and in Fig. 3.5—coefficient $m_z^{\delta_e}$.

From the above figures, it can be seen that even a two-fold error in setting the specified coefficients leads to an excess of the permissible angle of attack by only 5

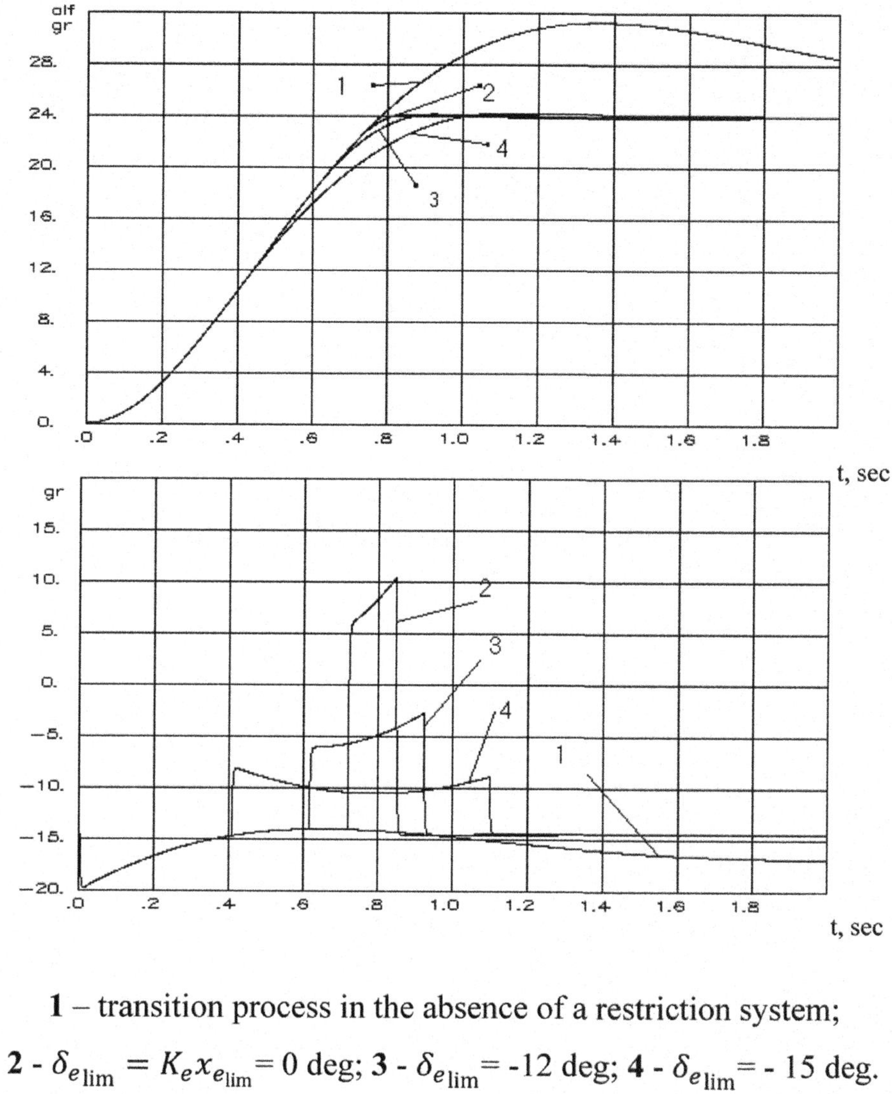

Fig. 3.2 Results of solving the model problem of limiting the angle of attack with different initial settings of the limiting system and ideal steering drives. **1**—transition process in the absence of a restriction system; **2**—$\delta_{e_{\mathrm{lim}}} = K_e x_{e_{\mathrm{lim}}} = 0$ deg; **3** $\delta_{e_{\mathrm{lim}}} = -12$ deg; **4**—$\delta_{e_{\mathrm{lim}}} = -15$ deg

… 8%. Errors in setting other coefficients have an even smaller impact. This indicates the robustness of the algorithm for limiting the angle of attack to modeling errors.

Obviously, the traditional scheme of the algorithm of restriction of the angle of attack does not have the demonstrated capabilities. Its use is effective only in a narrow range of tuning modes. Thus, the received algorithm for adaptive restriction of the

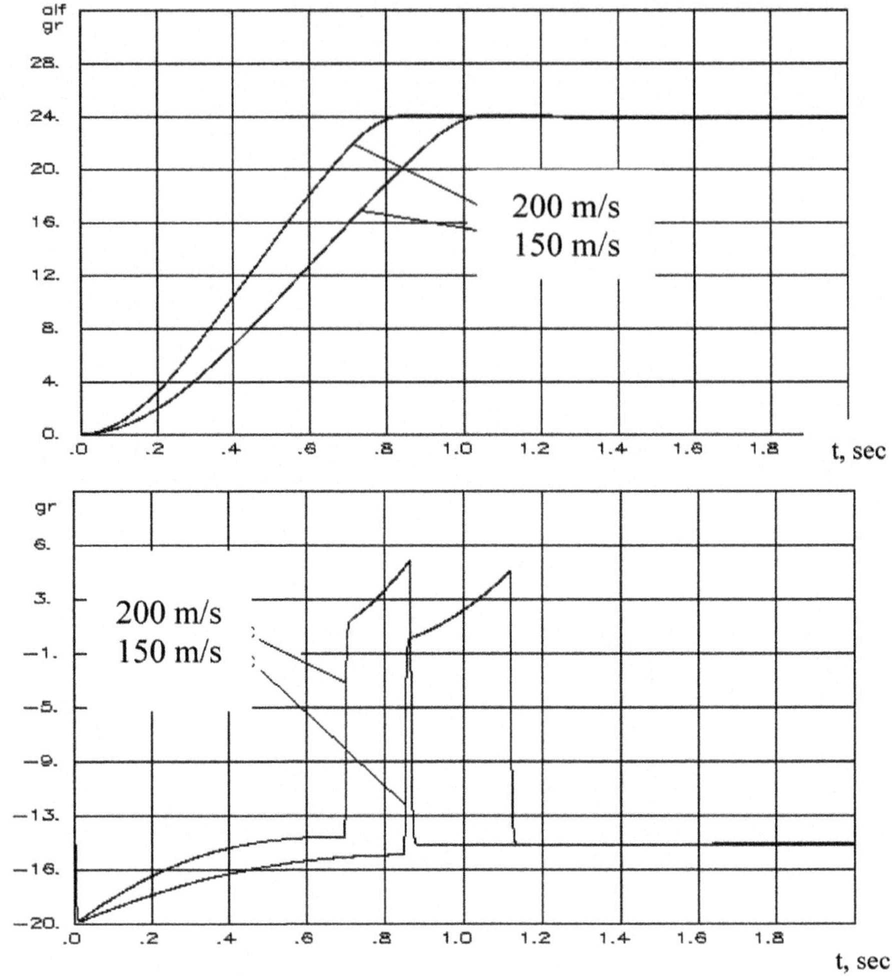

Fig. 3.3 Results of modeling the process of limiting the angle of attack at different initial flight speeds

angle of attack has clear advantages over similar traditional schemes for maintaining the angle of attack.

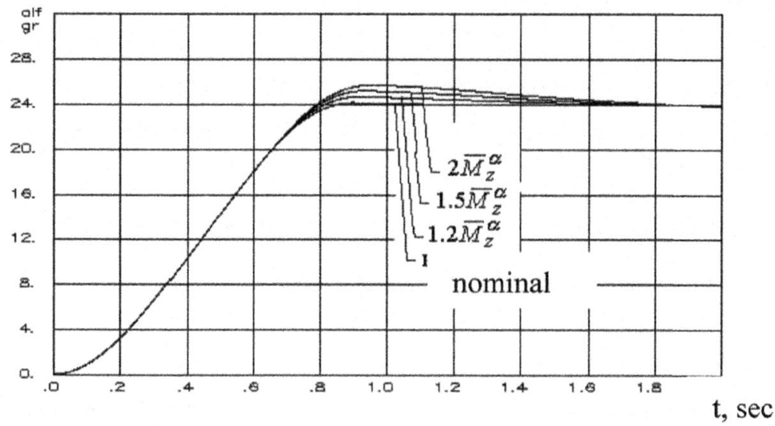

Fig. 3.4 Results of modeling the process of limiting the angle of attack at different values of the aerodynamic coefficient m_z^{α}

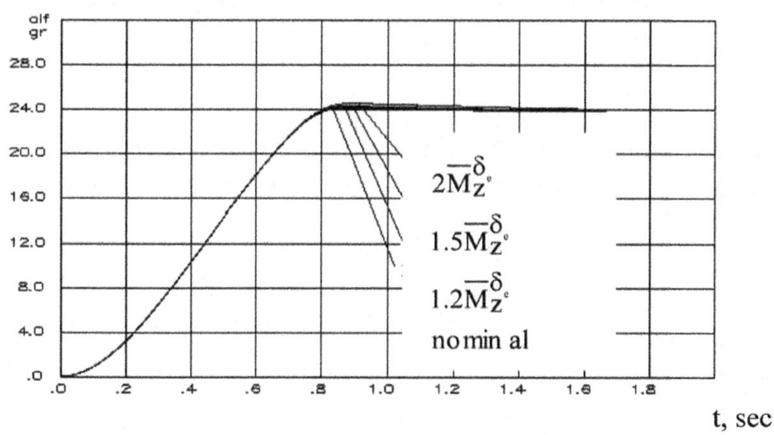

Fig. 3.5 Results of modeling the process of limiting the angle of attack at different values of the aerodynamic coefficient $m_z^{\delta_e}$

3.2 Peculiarities of Maintaining Restrictions on the Observed Parameters of Aircraft Movement

It should be noted that a large range of practical problems in the study of dynamic systems is based on the measurement of components that are not included in the state vector by means of a complex of sensors.

This section discusses the features of the method of maintaining restrictions on the components of the observation vector of a dynamic system by the example of limiting the normal overload of an aircraft. This may occur in an emergency situation, associated with an emergency landing of the AC, a lightning strike, etc. At the same time, to simplify the problem, an assumption about a full-component, instantaneous measurement of the needed parameters is made.

Let the motion of the control object be described by a differential equation of the form:

$$\dot{x} = Ax + Bu, \tag{3.12}$$

and the vector of observations is represented by the expression:

$$y = Cx, \tag{3.13}$$

where y—n-dimensional output vector, defined in space Y^n and C—matrix of constant coefficients of the corresponding dimension.

Write down the Eqs. (3.12) and (3.13) in the form of «input–output». To do this, we express the vector x from (3.13) and substitute the resulting expression in (3.12). After performing the corresponding transformations, we finally have

$$\dot{y} = CAC^{-1}y + CBu. \tag{3.14}$$

Note that in (3.14) and further on, if there are features of matrix C, the matrix inversion sign should be replaced by the pseudo-inversion sign.

We define a boundary hyperplane in the space Y^n by the following equation:

$$ay + c = 0, \tag{3.15}$$

where a—matrix-string of the corresponding dimension, c—scalar, and we obtain for the dynamic system, represented by the mathematical model (3.14), the basic relations of the control algorithm (2.20) of the adaptive system for maintaining constraints on the components of the vector y. In this case, we assume that the space Y^n is pre-normalized.

So, in accordance with the above methodology, the weighted distance to the boundary (3.15) is determined by the following equation:

$$d = \overline{a}y + \overline{c}. \tag{3.16}$$

Differentiating the expression (3.16) in time allows you to get the following dependencies:

$$\dot{d} = \overline{a}\left(CAC^{-1}y + CBu\right); \tag{3.17}$$

$$\ddot{d}_a = \overline{a}CA\left(AC^{-1}y + Bu_{\text{lim}}\right). \tag{3.18}$$

Ensuring $\ddot{d} = const$ at the restriction stage is carried out by adjusting the control of the restriction by the value

$$\Delta u(t) = (\overline{a}CAB)^{\#}\Delta\ddot{d}(t) = (\overline{a}CAB)^{\#}\left(\ddot{d}_a(t) - \ddot{d}_{a_{mean}}\right) \tag{3.19}$$

or taking into account the corresponding transformations:

$$\Delta u(t) = B^{\#}AC^{-1}(y(t) - y_{mean}). \tag{3.20}$$

In (3.20), y_{mean}—the vector of observations at the time of activation of the restriction system. From the point of view of computational difficulties, the form of writing (3.19) is more convenient, since under the sign of the pseudo-circulation there is a matrix-string, and as a result, this procedure is solved analytically.

We apply the approach, described here, to the problem of limiting normal overload.

Let the motion of the AC be described by a system of linear differential equations of the second order:

$$\underbrace{\begin{bmatrix} \dot{\alpha} \\ \dot{\omega}_z \end{bmatrix}}_{\dot{x}} = \underbrace{\begin{bmatrix} -\overline{Y}^{\alpha} & 1 \\ \overline{M}_z^{\alpha} & \overline{M}_z^{\omega_z} \end{bmatrix}}_{A} \underbrace{\begin{bmatrix} \alpha \\ \omega_z \end{bmatrix}}_{x} + \underbrace{\begin{bmatrix} 0 \\ \overline{M}_z^{\delta_e} \end{bmatrix}}_{B} \underbrace{[\delta_e]}_{u}. \tag{3.21}$$

We will consider the nmal overload n_y and angular velocity ω_z as the observed parameters. In this case, the expression (3.13) is rewritten as follows:

$$\underbrace{\begin{bmatrix} n_y \\ \omega_z \end{bmatrix}}_{y} = \underbrace{\begin{bmatrix} n_y^{\alpha} & 0 \\ 0 & 1 \end{bmatrix}}_{c} \underbrace{\begin{bmatrix} \alpha \\ \omega_z \end{bmatrix}}_{x}. \tag{3.22}$$

Then, given in the space Y by the constraint $n_{y_{\lim}}$ in the form, shown in Fig. 3.6 (that is, by setting the parameters of the constraint line (3.15) in the form: $a = [-1\ 0]$, $c = n_{y_{\lim}}$) we get the expanded expressions (3.16)–(3.19).

So,

$$d = n_{y_{\lim}} - n_y; \tag{3.23}$$

$$\dot{d} = \overline{Y}^{\alpha} n_y - n_y^{\alpha}\omega_z; \tag{3.24}$$

$$\ddot{d}_a = \left(\overline{Y}^{\alpha} - \overline{M}_z^{\omega_z}\right) n_y^{\alpha}\omega_z - \left(\left(\overline{Y}^{\alpha}\right)^2 + \overline{M}_z^{\alpha}\right) n_y - \overline{M}_z^{\delta_e} n_y^{\alpha}\delta_{e_{\lim}}; \tag{3.25}$$

$$\Delta\delta_e(t) = -\frac{\Delta\ddot{d}(t)}{n_y^{\alpha}\overline{M}_z^{\delta_e}}. \tag{3.26}$$

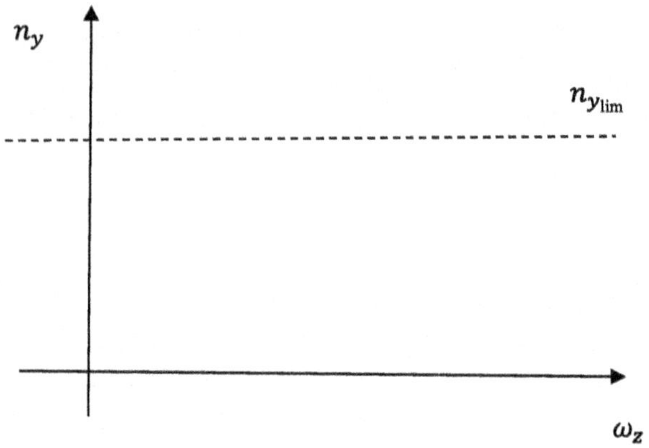

Fig. 3.6 Type of restriction $n_{y\lim}$

Equations (3.23)–(3.26) are the main dependencies of the algorithm for the functioning of the adaptive system for limiting normal overload when using the available information about the parameters of the aircraft movement. Figure 3.7 presents the transition processes of the aircraft's exit to overload under conditions of restrictions, both under ideal conditions and taking into account the flight information sensors.

The book used a model of sensors in the form of a second-order dynamic system with error generators in the following form [7]:

$$\begin{aligned}\dot{y}_1 &= y_2;\\ \dot{y}_2 &= -\omega^2 y_1 - 2\xi\omega y_2 + \omega^2 x + \eta;\\ z &= ky_1 + \varsigma\end{aligned} \tag{3.27}$$

where y_1, y_2—auxiliary variables; x—useful, measurable signal; ω—frequency of own underfed oscillations of the measuring element of the sensor; ξ—relative vibration damping coefficient of the measuring element of the sensor; η—interference at the sensor input; k—coefficient of sensor transmission; ζ—sensor errors, that occur, when the signal is picked up.

As can be seen from the simulation results, the algorithm for limiting the normal overload is robust to the errors of real sensors.

In Fig. 3.8, the results of the evaluation of the influence of the time constant of the control system on the accuracy of maintaining the limit $n_{y\lim}$ are presented. In this case, the dynamic properties of the control system were described by a differential equation of the form:

$$T_{\text{sd}}\Delta\dot{y} + \Delta y = k_g \Delta z, \tag{3.28}$$

where T_{sd}—steering drive time constant; k_g—coefficient of gain; Δy, $\Delta\dot{y}$—accordingly, the position and speed of the output rod of the steering drive; Δz—moving the steering spool.

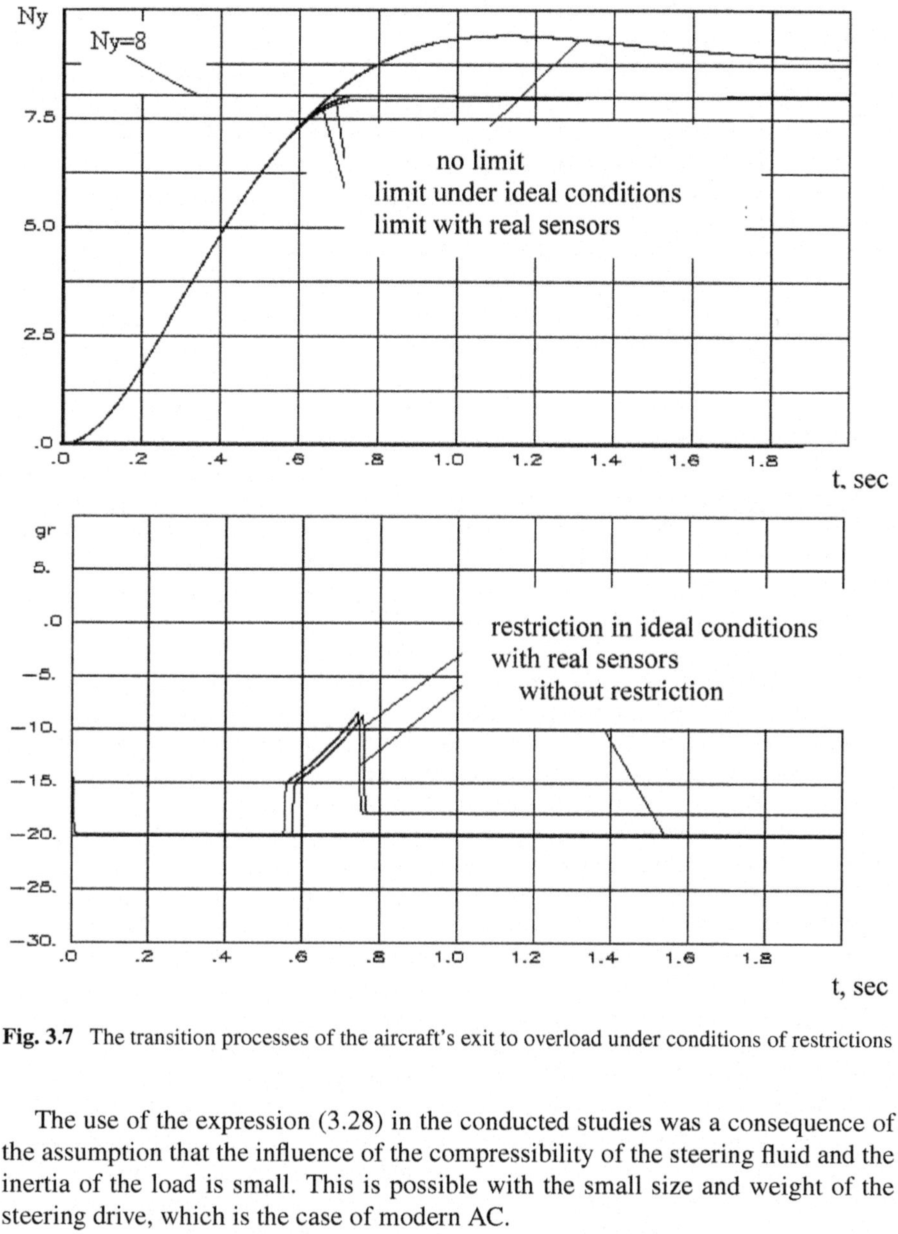

Fig. 3.7 The transition processes of the aircraft's exit to overload under conditions of restrictions

The use of the expression (3.28) in the conducted studies was a consequence of the assumption that the influence of the compressibility of the steering fluid and the inertia of the load is small. This is possible with the small size and weight of the steering drive, which is the case of modern AC.

The analysis of the obtained results shows that the accuracy of maintaining the limit $n_{y_{\lim}}$ naturally depends on the speed of the steering drive.

However, the simulation of the restriction process with the real characteristics of the steering drive (Tsd = 0.05) gives an acceptable accuracy of the output of the aircraft on $n_{y_{\lim}}$. The error is approximately 0.3 overload units, which corresponds

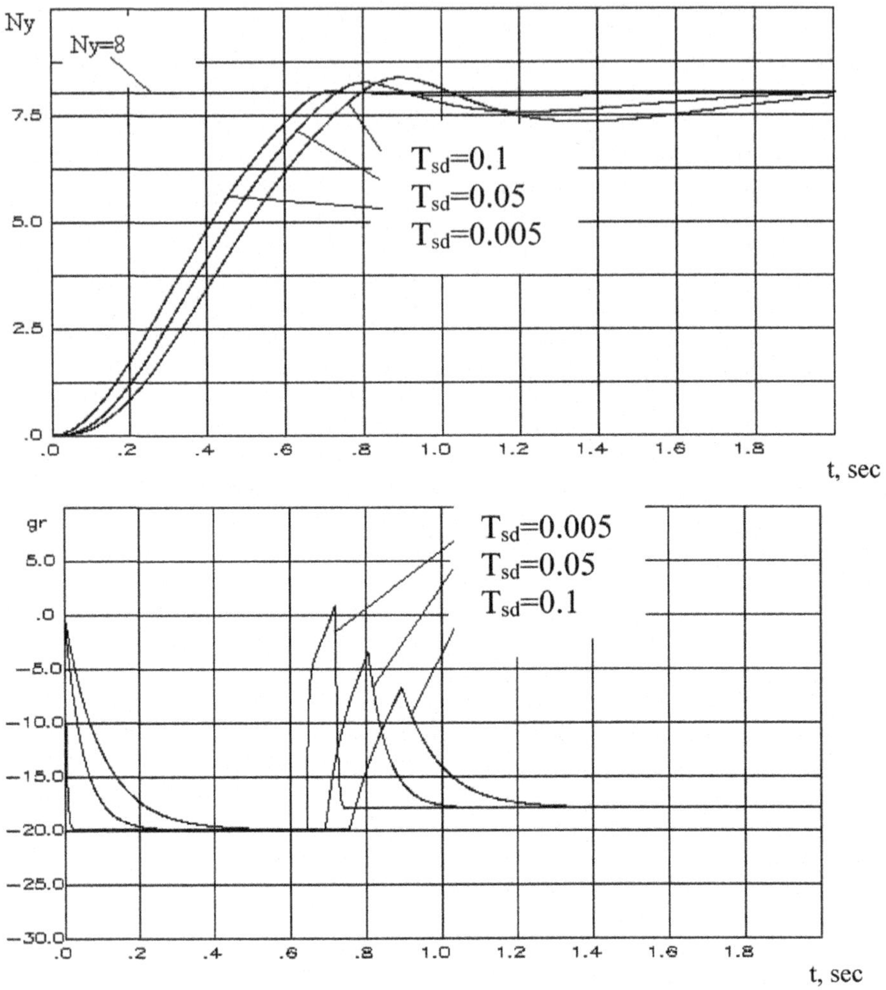

Fig. 3.8 Results of the evaluation of the influence of the time constant of the control system on the accuracy of maintaining the limit $n_{y\text{lim}}$

to 4% of the allowed value. Thus, we can draw a conclusion about the robustness of the restriction algorithms and the real characteristics of the steering drives.

3.3 Limiting the Roll Angle of the Helicopter

By itself, the tilt of the helicopter is not fraught with danger. However, the roll of the helicopter, from the conditions of ensuring the safety of flight at low altitudes, is limited to a value proportional to the current altitude. That is, the allowed roll in degrees corresponds to the number of meters of height.

The helicopter, as an object of control, has a number of unfavorable characteristics of stability and controllability, the main of which are as follows:

- small damping moments for angular velocities of pitch, roll, and yaw;
- static instability of the main rotor in the angle of attack;
- if there is static stability in pitch, roll, and speed of flight, the helicopter does not have dynamic stability in pitch and speed in all flight modes, and in roll-in hovering and low-speed flight modes;
- the relationship between roll and yaw movements in progressive flight, which leads to oscillatory instability of lateral motion, when flying at low speeds, and too weakly attenuated oscillations at high speeds.

Thus, the helicopter, as a control object, is a rather complex dynamic system [8]. In this case, the control of its angular position only by pitch, roll, or yaw leads to the need to move the command control levers along all the listed coordinates simultaneously. In this regard, to obtain the main dependencies of the algorithm for maintaining the maximum roll of the helicopter, it is necessary to consider at least a five-dimensional state space X, the state vector of which includes the following parameters [9, 10]:

$$x = \left[\gamma\ \vartheta\ \omega_x\ \omega_y\ \omega_z\right]^m \tag{3.29}$$

and the vector of controls δ, including

$$\delta = \left[\delta_x\ \delta_z\ \phi_{\text{gp}}\right]^m, \tag{3.30}$$

where δ_x, δ_z—the deviation of the cyclic pitch handle in roll and pitch, respectively; ϕ_{gp}—deviation of the general pitch handle. Let's use the mathematical model of the helicopter, presented in the following form:

$$\begin{aligned}
\dot{\gamma} &= \omega_x - tg\vartheta\left(\omega_y \cos\gamma - \omega_z \sin\gamma\right);\\
\dot{\vartheta} &= \omega_z \cos\gamma + \omega_y \sin\gamma;\\
\dot{\omega}_x &= \frac{M_x}{I_x} + \frac{I_z - I_y}{I_x}\omega_y\omega_z;\\
\dot{\omega}_y &= \frac{M_y}{I_y} + \frac{I_x - I_z}{I_y}\omega_x\omega_z;\\
\dot{\omega}_z &= \frac{M_z}{I_z} + \frac{I_y - I_x}{I_z}\omega_x\omega_y.
\end{aligned} \tag{3.31}$$

In (3.31), the moments, acting on the helicopter, are determined in accordance with the force diagram, shown in Fig. 3.9.

Thus,

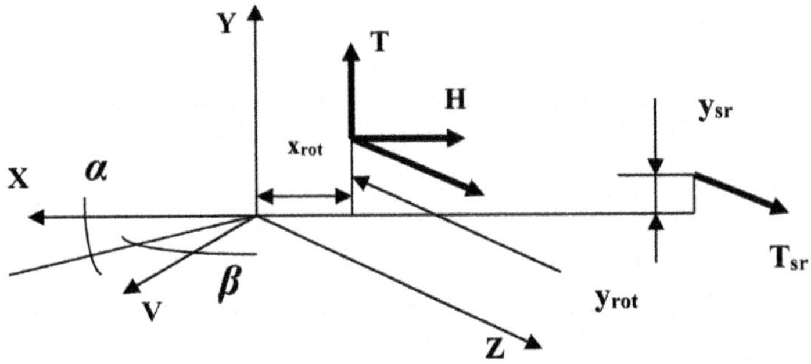

Fig. 3.9 Diagram of the forces, acting on the helicopter

$$\begin{aligned} M_x &= -I_\omega \omega_{\text{rot}} \omega_z + \left(S y_{\text{rot}} + M_{x_{\text{gl}}} + M_{x_{\text{rot}}} + M_{x_{\text{sr}}}\right) \\ M_y &= S x_{\text{rot}} + M_{x_{\text{gl}}} + M_{x_{\text{rot}}} + M_{x_{\text{sr}}} \\ M_z &= I_\omega \omega_{\text{rot}} \omega_x + \left(-T x_{\text{rot}} + H y_{\text{rot}} + M_{z_{\text{gl}}} + M_{z_{\text{rot}}} + M_{z_{\text{sr}}}\right) \end{aligned} \tag{3.32}$$

where T, H, S—projections of the resulting force of the main rotor on the associated axes; $M_{x_{\text{rot}}}, M_{y_{\text{rot}}}, M_{z_{\text{rot}}}$—projections of the resulting force of the main rotor; $M_{x_{\text{gl}}}, M_{y_{\text{gl}}}, M_{z_{\text{gl}}}$—projections of the resulting force of the glider; $M_{x_{\text{sr}}}, M_{y_{\text{sr}}}, M_{z_{\text{sr}}}$—projections of the resulting force of the steering rotor; $x_{\text{rot}}, y_{\text{rot}}$—the position of the main rotor hub relative to the center of mass of the helicopter; I_ω—the moment of inertia of the main rotor blades relative to the horizontal hinges; ω_{rot}—angular velocity of the main rotor.

The main rotor forces and torques are given as follows [11]:

$$\begin{aligned} T &= T_0 - T_0^{\omega_x} \omega_x + 4T_0^{\omega_z} \omega_z \\ H &= H_0 - H_0^{\omega_x} \omega_x + 4H_0^{\omega_z} \omega_z + T(D_1 \delta_z + D_2 \delta_x) \\ S &= S_0 - S_0^{\omega_z} \omega_z - S_0^{\omega_x} \omega_x + T(D_2 \delta_z - D_1 \delta_x) \\ M_{x_{\text{rot}}} &= M_{x_{\text{rot0}}} + 0,5 I_\omega \omega_{\text{rot}} (D_2 \delta_z - D_1 \delta_x) \\ M_{z_{\text{rot}}} &= M_{z_{\text{rot0}}} + 0,5 I_\omega \omega_{\text{rot}} (D_1 \delta_z + D_2 \delta_x); \end{aligned}$$

$M_{y_{\text{rot}}}$ is set in a table depending on the total pitch of the main rotor; D_1, D_2—kinematic parameters of the hub.

Given the control algorithm of the restriction system of roll angle for helicopter, in relation to the deviation of the skew automat in the roll channel, we have the following:

$$\delta_x = \begin{cases} \delta_x^p + k_k^\gamma (\gamma - \gamma_g) + k_k^{\omega_x} \omega_x, & d > d_{\text{lim}}; \\ \delta_x^p + \Delta\delta_x(t) + k_k^{\omega_x} \omega_x, & d > d_{\text{lim}}; \end{cases} \tag{3.33}$$

In the pitch channel, the ACS law of angular position stabilization is worked out. Using the above method for the general case of restriction, we obtain the main relations of the algorithm.

As a result [12],

- current distance to the acceptable roll angle value $d = \gamma_{\text{lim}} - \gamma$;
- the first derivative of the distance d:

$$\dot{d} = -\omega_x + tg\vartheta\left(\omega_y \cos\gamma - \omega_z \sin\gamma\right);$$

- available braking acceleration, when control δ_x^{lim} is available:

$$\ddot{d}_a = -\frac{M_x\left(\delta_x^{\text{lim}}\right)}{I_x} + tg\vartheta\left(\frac{M_y\left(\delta_x^{\text{lim}}\right)}{I_y}\cos\gamma - \frac{M_z\left(\delta_x^{\text{lim}}\right)}{I_z}\sin\gamma\right),$$

  (this expression is obtained with the exception of summands of the second order of smallness from review);
- correction of control at the restriction stage $\Delta\delta_x(t) = \left[\frac{D_1(T y_{\text{rot}}+0,5 I_\omega \omega_{\text{rot}})}{I_x} - tg\vartheta\left(\frac{T D_1 x_{\text{rot}}}{I_y}\cos\gamma + \frac{D_2(T y_{\text{rot}}+0,5 I_\omega \omega_{\text{rot}})}{I_z}\sin\gamma\right)\right]^{\#}\Delta\ddot{d}(t)$.

The results of modeling the restriction of the helicopter roll angle are presented in Figs. 3.10 and 3.11.

Figure 3.10 illustrates the helicopter's exit at a roll angle of 25° and its restriction at the level of 15°. The slight over-regulation, noticeable in the figure, is explained by taking into account the real properties of the steering drive of the deflection plate of the skew automat in the simulation.

Figure 3.11 characterizes the adaptive properties of the algorithm to change the flight mode for a similar constraint problem.

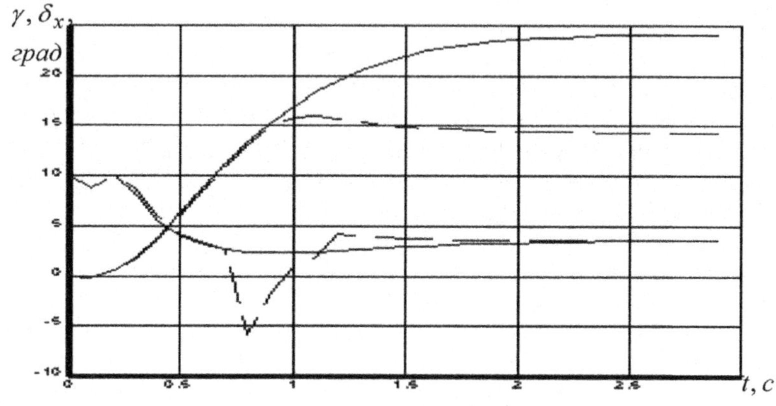

Fig. 3.10 The helicopter's exit at a roll angle of 25° and its restriction at the level of 15°

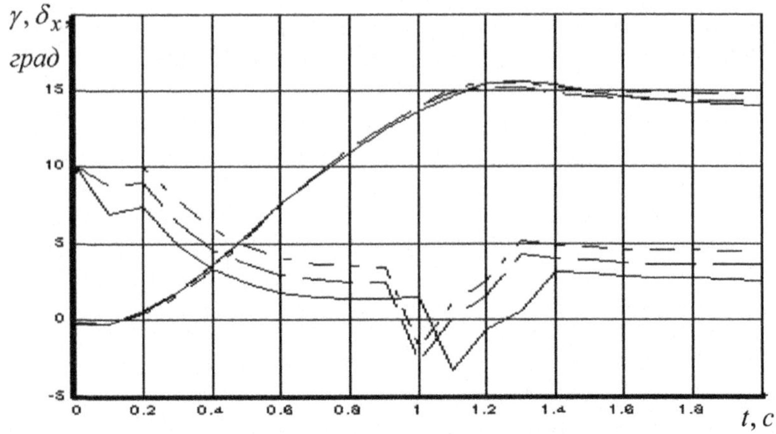

Fig. 3.11 The adaptive properties of the algorithm to change the flight mode for a similar constraint problem

3.4 Restriction in Space of Two Components of the State Vector by the Example of the Problem of Limiting the Angles of Attack and Slip of an Aircraft

In the previous paragraphs, the problems of limiting one of the components of the state vector of the aircraft, as well as the values calculated on their basis, are considered. This section deals with the problem of maintaining two-dimensional, interconnected constraints.

Consider the model problem of maintaining the boundaries of the operational area in the cross section (α, β) for AC, the motion of which is described by a system of linear differential equations of the fifth order [13–15]:

$$\dot{x} = Ax + Bu, \tag{3.34}$$

where $x = [\alpha\ \beta\ \omega_x\ \omega_y\ \omega_z]^m$; $u = [\delta_e\ \delta_n]^m$.

The boundary of the operational area is defined by the cross section of the space X in coordinates (α, β) in the form, shown in Fig. 3.12, with a dashed line.

The solid line characterizes the standard setting for a one-parameter system of restriction of the angle of attack.

We obtain the calculated dependences of the algorithm for maintaining the constraint for the boundary, given by the equation

$$ax + c = 0 \tag{3.35}$$

and caused by positive values of the sliding angle β. In accordance with the above figure, the parameters of the equation of the constraint line (3.35) are given as

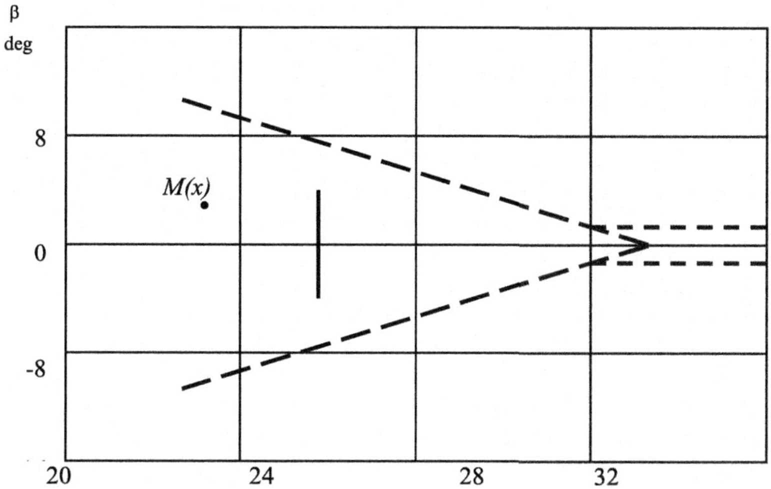

Fig. 3.12 View of the operational area boundary

$$a = [-1 \ -1\, 0\, 0\, 0], \quad c = 34.$$

Then the shortest distance from an arbitrary point *M(x)* that characterizes the state of the constraint object in space *X* to the boundary (3.35) is defined as follows:

$$d = \overline{a}x + \overline{c}, \tag{3.36}$$

where $\overline{a} = \frac{a}{|a|} = \left[-\sqrt{2}/2 - \sqrt{2}/2\ 0\ 0\ 0\right]$, $\overline{c} = \frac{c}{|a|} = 17\sqrt{2}$.

In accordance with the above methodology, for this description of the motion of the aircraft by the system (3.34), the equations of the synthesis of the system of maintaining constraints, considered in relation to a deterministic linear continuous dynamic system in the stationary case, are applicable. Namely,

$$u = \begin{cases} u(t), & d > d_{\lim}; \\ u_{\lim} + \Delta u(t), & d \le d_{\lim}, \end{cases}$$

where

$$d_{\lim} = \frac{\dot{d}^2}{2\ddot{d}_a};\quad \dot{d} = \overline{a}(Ax + Bu);\quad \ddot{d}_a = \overline{a}A(Ax + Bu_{\lim});\quad \Delta u(t) = (\overline{a}AB)^{\#}\Delta\ddot{d}(t).$$

By setting the constraint control in the form:

$$u_{\lim} = \left[\delta_{ds_{\lim}}\ \delta_{dr_{\lim}}\right]' \tag{3.37}$$

where $\delta_{ds_{\text{lim}}}$—the specified available deviation of the stabilizer; $\delta_{dr_{\text{lim}}}$—given the available deviation of the rudder to reduce the slip, we can write the expanded expressions of the components of the constraint algorithm:

- distance to the restriction boundary:

$$d = -\frac{1}{\sqrt{2}}(\alpha + \beta - 34); \tag{3.38}$$

- the first derivative of the distance:

$$\dot{d} = \frac{1}{\sqrt{2}}\left(\overline{Y}^{\alpha}\alpha - \overline{Z}^{\beta}\beta - \omega_y - \omega_z\right); \tag{3.39}$$

- available braking acceleration:

$$\ddot{d}_a = -\frac{1}{\sqrt{2}}\left(\begin{array}{l} \overline{Y}^{\alpha^2} + \overline{Z}^{\beta^2} + \overline{M}_z^{\alpha}\alpha + \overline{M}_y^{\beta}\beta + \left(\overline{M}_z^{\omega_z} - \overline{Y}^{\alpha}\right)\omega_z \\ + \left(\overline{M}_y^{\omega_y} + \overline{Z}^{\beta}\right)\omega_y + \overline{M}_y^{\delta_n}\delta_{dr_{\text{lim}}} + \overline{M}_z^{\delta_e}\delta_{ds_{\text{lim}}} \end{array}\right); \tag{3.40}$$

- correction of control at the restriction stage:

$$\Delta u(t) = \left[-\frac{\overline{M}_y^{\delta_n}}{\sqrt{2}} \; -\frac{\overline{M}_y^{\delta_e}}{\sqrt{2}}\right]^{\#} \Delta\ddot{d}(t). \tag{3.41}$$

The operation of the algorithm for maintaining the constraint in coordinates (α, β) illustrates in Figs. 3.13, and 3.14.

In the simulation, the constraint problem was formulated as follows: to ensure the non-plugging of the constraint line in the coordinates (α, β). The task of balancing the object at the level of acceptable values of parameters was not considered.

The analysis of the simulation results shows a high efficiency in maintaining the restriction. In addition, we can conclude that the method is adaptable to flight modes. So, reducing the speed of the AC flight only changes the point of time, when the algorithm of the restriction is turned on.

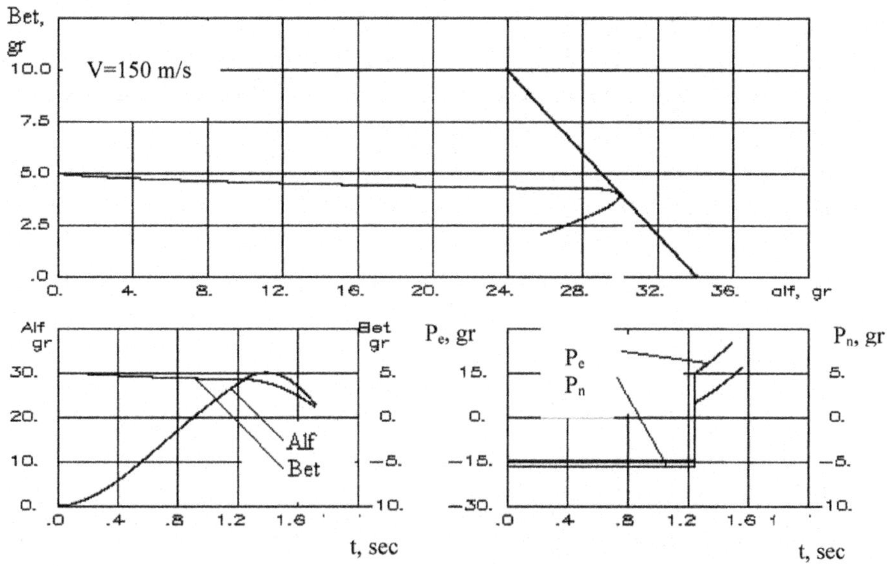

Fig. 3.13 The operation of the algorithm of constraint maintaining for V=150 m/s

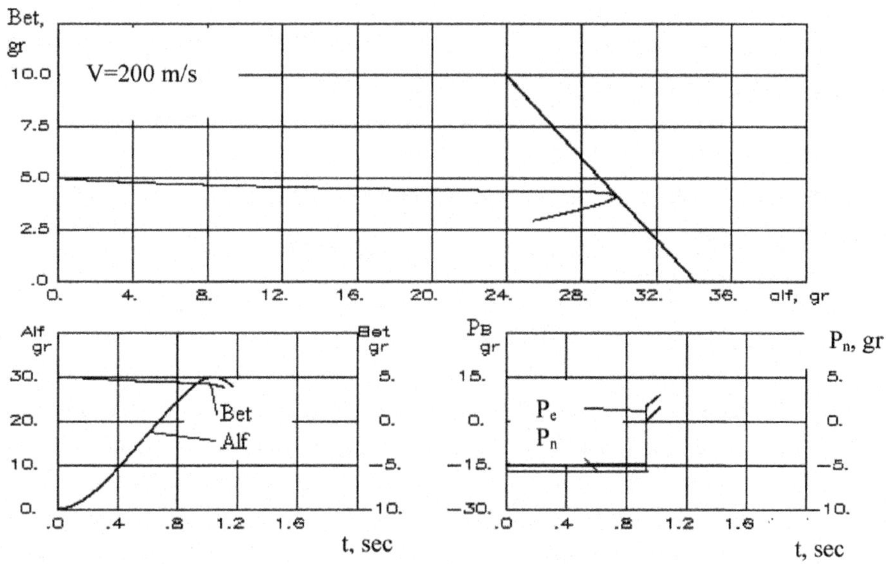

Fig. 3.14 The operation of the algorithm of constraint maintaining for V=200 m/s

References

1. Ivanov VS (ed) (2003) Flight safety of aircraft. - M.: Air Force Engineering Academy named by N.E. Zhukovsky
2. Starikov AN (ed) (1988) Flight safety of aircraft: Methodological foundations. -M.: Transport
3. Proctor P (1999) Helmet displays boost safety and lethality. In: Aviation week & space technology. pp 81–82
4. Proctor P (1998) Integrated cockpit safety system certified. In: Aviation week & space technology. No 14, p 61
5. Akimov AN (1992) Method of identification of failures of dynamic systems. In: Avtomatika i telemekhanika. No 6
6. Vorobyov VV (2000) Modified method for identifying failures in dynamic systems. In: Automation and telemechanics. No 11
7. Kochetkov YA, Kulifeev YB, Podobedov VA (1978) Mathematical models of sensors. In: Research on aeroautoelasticity.– M.: Proceedings of air force engineering academy named by N.E. Zhukovsky
8. Geoffrey JJ (2002) Open design for helicopter active control systems. American Helicopter Society 58th Annual Forum, Montreal, Canada
9. Krasovsky AA (1996) Attractors and synthesis of controls in critical modes. In: Izvestiya akademii nauk. Theory and control systems, No 3
10. Krasovsky AA (1987) Handbook of the theory of automatic control. - M.: Nauka, The main editorial office of physical and mathematical literature
11. Pogrebnyak VI (2001) Accident rate in helicopter aviation. In: Problems of flight safety. - M.: VINITI
12. Zatuchny DA (2019) Approach to determining the reliability of the navigation information, transmitted from the helicopter. In: Proceedings of the international symposium «Reliability and Quality», Penza, vol 1, pp 217–219
13. William B/Scott/Edwards AFB, Calif (1999) Automatic GCAS: «You can't fly any lower». In: Aviation week & space technology. p 76–80
14. Ukolov IS, Kirsanov BV, Noskov YV (1981) Ensuring flight safety with the help of an active control system/issues of cybernetics. In: Problems of aviation and space cybernetics. - M.: Scientific Council of the USSR Academy of Sciences on the space problem «Cybernetics»
15. Mayorov AV, Moskatov GK, Shibanov GP (1988) Safety of functioning of automated objects. - M.: Mashinostroenie

Chapter 4
Algorithms for Adaptive Limitation of Trajectory Parameters of Aircraft Movement

4.1 General Case of Limiting the Trajectory of Aircraft Movement in Three-Dimensional State Space

4.1.1 Substantial Problem Statement

Let be in the normal coordinate system $OX_gY_gZ_g$ are determined (see Fig. 4.1): the convex constraint surface, given by the equation:

$$\xi(x) = 0, \tag{4.1}$$

point $M(x_0)$ with the current coordinates of the aircraft, obtained either using a digital map of the area, or by methods of accurate navigation (these issues are not discussed in this book), the values of the flight parameters, measured on board the AC, are known.

Required: in the event of a collision of the AC with the restriction surface, approximated by Eq. (4.1), ensure that the aircraft evades it. The control of the AC in the process of withdrawal is carried out in the pitch, roll, and course channels. In this case, it is assumed that the control commands are generated by the restriction system within the framework of the control, entrusted to it.

4.1.2 Object Model and Constraint Surfaces

Let's define a model of the spatial motion of the AC in the form:

$$\begin{cases} \dot{x} = f(w); \\ \dot{w} = \phi(w, u), \end{cases} \tag{4.2}$$

V. Akimov and V. A. Zatuchny, *Limiting Modes of Aircraft Flight*,
Springer Aerospace Technology, https://doi.org/10.1007/978-981-19-6329-2_4

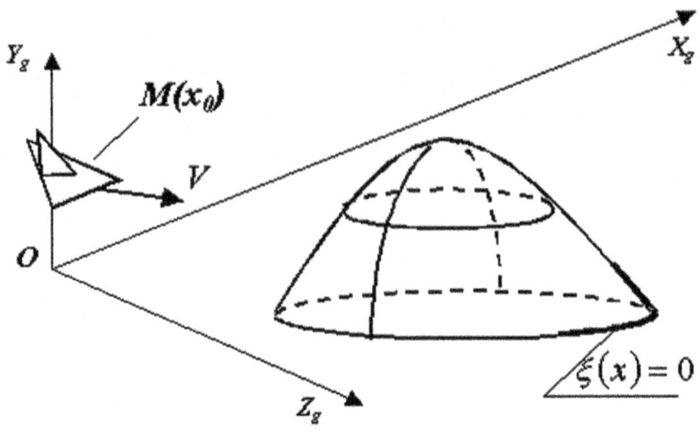

Fig. 4.1 Position of aircraft and surface restrictions

where $x = [X_g Y_g Z_g]^T$—vector that characterizes the position of the center of mass of the aircraft in the normal coordinate system; $w = [V\theta\psi]^T$—vector that defines the module and orientation of the AC velocity relative to the normal coordinate system $OX_gY_gZ_g$; $u = [n_x n_y \gamma]^m$—vector of control; $f(w)$, $\phi(w, u)$—vector functions of the specified vector arguments.

Note that in the system (4.2), the control u has a direct effect on the vector of parameters w, and on the vector x—the influence is made through the vector w, that is, the vector w is a pseudo-control for the vector x, the set of parameters of which are subject to restriction.

We will entrust control, defined in some way, to the restriction system:

$$u_{\lim} = [n_{x_{\lim}} \, n_{y_{\lim}} \, \gamma_{\lim}]^m \tag{4.3}$$

and use the method of synthesis of its algorithms, described for a deterministic continuous controlled multilevel dynamical system (see Chap. 2).

To simplify the calculations, we will use the above approach to describe the boundary of the constraint surface. Namely, at the point of the predicted collision of the AC with the constraint surface, a tangent plane is constructed, which is taken as the constraint plane. This approach is acceptable due to the cyclic operation of the constraint algorithm.

If we limit ourselves to considering the linear forecast of the trajectory of the aircraft, then the coordinates of the point of the predicted collision T (see Fig. 4.2) are determined by jointly solving the equations of the constraint surface (4.1) and the line, containing the velocity vector of the aircraft.

The canonical representation of the latter has the form:

$$\frac{X_g - X_{g0}}{V_x} = \frac{Y_g - Y_{g0}}{V_y} = \frac{Z_g - Z_{g0}}{V_z}, \tag{4.4}$$

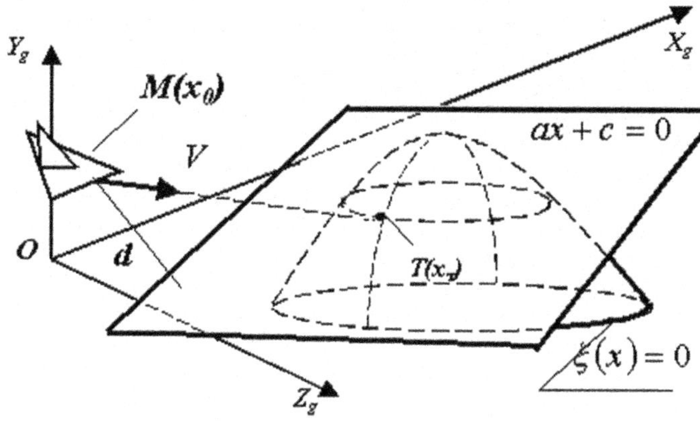

Fig. 4.2 Model of the predicted collision of the AC with the constraint surface

where V_x, V_y, V_z—projections of the velocity vector on the axis of the normal coordinate system.

Then, according to [1], the plane, tangent to the surface of the constraint at the point $T(x_T)$, can be represented by the equation:

$$ax + c = 0, \tag{4.5}$$

where

$$a = \frac{\partial \xi(x_m)}{\partial x^m}; \; C = -\frac{\partial \xi(x_m)}{\partial x^m} x_m$$

When the AC evolves in space, due to the control actions of the pilot, ACS, or the restriction system, the coordinates of the point $T(x_m) \in \xi(x)$ change. This, in turn leads to the need for cyclic recalculation of the parameters of the constraint plane, defined by the matrix-string a and the scalar c [2].

4.1.3 Derivation of the Calculated Dependencies of the Limiting Algorithm

By analogy with the conclusion of the calculated dependencies (2.45–2.48) of the algorithm for maintaining constraints on the parameters of the state vector of a deterministic continuous controlled multilevel dynamical system for the case under consideration, we have.

- the shortest distance from the AC to the constraint plane:

$$d = \overline{a}x + \overline{c}; \tag{4.6}$$

- the speed at which the AC approaches the constraint surface (the first derivative of the distance):

$$\dot{d} = \bar{a}\dot{x} = \bar{a} f(w); \tag{4.7}$$

- available acceleration of the AC braking system, when the restriction system is activated:

$$\ddot{d}_a = \overline{a}\frac{d}{dt} f(w) = \overline{a}\frac{\partial f(w)}{\partial w^m}\phi(w, u_{\text{lim}}); \tag{4.8}$$

- control correction at the restriction stage:

$$\Delta u(t) = \left(\overline{a}\frac{\partial f(w)}{\partial w^m}\,\frac{\partial \phi(w, u)}{\partial u^m}\right)^{\#} \Delta\ddot{d}(t). \tag{4.9}$$

In (4.8–4.9), the matrices $\frac{\partial f(w)}{\partial w^T}$, $\frac{\partial \phi(w,\,u)}{\partial u^T}$ are obtained by formally differentiating the vector functions *f(w)*, *ϕ(w, u)* with respect to *w* and *u*, respectively, as a parameter. In accordance with this, the matrices of partial derivatives have the form:

$$\frac{\partial f(w)}{\partial w^m} = \begin{bmatrix} \cos\,\theta \cos\psi & -V\sin\theta\cos\psi & -V\cos\theta\sin\psi \\ \sin\theta & V\cos\theta & 0 \\ \cos\theta\sin\psi & V\sin\theta\sin\psi & -V\cos\theta\cos\psi \end{bmatrix}; \tag{4.10}$$

$$\frac{\partial \phi(w, u)}{\partial u^m} = \begin{bmatrix} g & 0 & 0 \\ 0 & \frac{g\cos\gamma}{V} & -\frac{gn_y\sin_\gamma}{V} \\ 0 & \frac{g\sin\gamma}{V\cos\theta} & -\frac{gn_y\cos\gamma}{V\cos\theta} \end{bmatrix} \tag{4.11}$$

In Fig. 4.3, the trajectory of the AC withdrawal from the constraint surface in the form of a paraboloid of rotation is presented.

The restriction algorithm is activated, when the condition occurs:

$$d \le d_{lim} = \frac{\dot{d}^2}{2\ddot{d}_a}.$$

When the constraint algorithm is activated, under the control of $u_{\text{lim}} + \Delta u(t)$ AC changes its position in space. Therefore, the point $T(x_T)$ on the surface of the constraint $\xi(x) = 0$ (the family of points T on a paraboloid (see Fig. 4.3) makes up the line L) changes its position. Accordingly, the parameters of the constraint plane

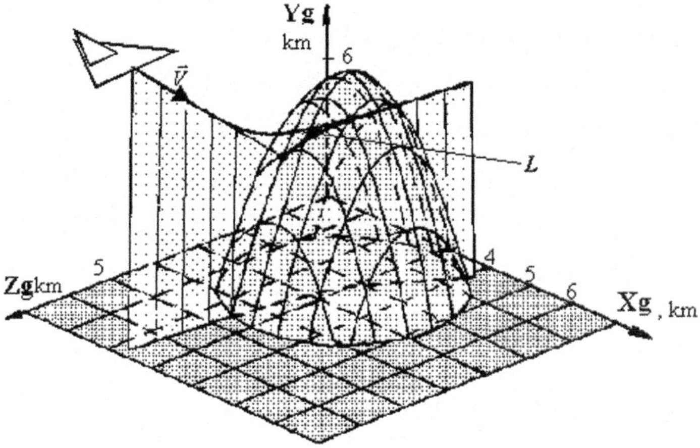

Fig. 4.3 The trajectory of the AC withdrawal from the constraint surface in the form of a paraboloid of rotation

(*a, c*) and the conditions for the algorithm's activation change. The restriction system is turned off, when the threat of a collision with the surface disappears.

A similar problem with similar results was solved by the authors of the monograph [3]. However, the method of direct optimization, used by them, is much more computationally complex. In addition, it excludes the AC crew from the process of forming the flight path. The latter is unacceptable, when solving tasks, aimed at ensuring the safety of a flight with a crew in the control circuit of the AC.

4.2 Particular Problems of Limiting the Trajectory Parameters of Movement in the Vertical and Horizontal Planes

4.2.1 Aircraft Drift in the Vertical Plane

Let us consider the problem of synthesis of algorithms for moving the AC in a vertical plane away from the constraint surface, given by an inclined line with the equation:

$$ax + c = 0. \tag{4.12}$$

The graphical interpretation of the problem in the coordinates H, L is shown in Fig. 4.4.

To describe the motion of the center of mass of the AC, we use the model (4.2) and methodically investigate two cases. Let us assume that in the first case the AC decreases at a constant rate, and in the second case $V = var$.

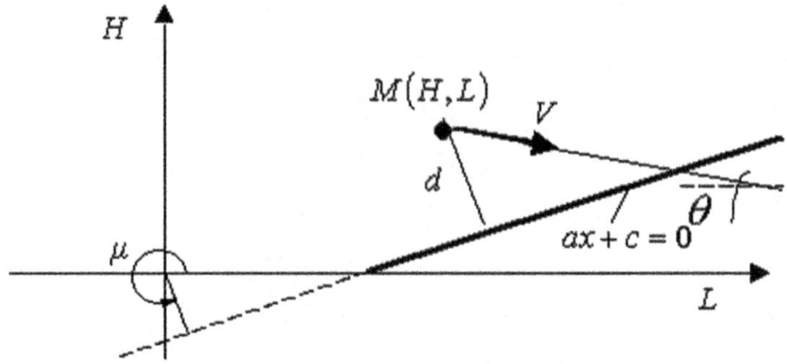

Fig. 4.4 Graphical interpretation of the problem of synthesis of algorithms for moving the aircraft in the vertical plane away from the constraint surface

1. Withdrawal of the aircraft with $V = const.$

For this case, the system (4.2) will have the following components. State vector $x = [H \;\; L]^T$, pseudo-control vector $w = [\theta]$, and control vector $u = [n_y]$. This corresponds to the system of equations of the third order of motion of the center of mass of the AC in the vertical plane:

$$\begin{cases} \dot{H} = V \sin\theta; \\ \dot{L} = V \cos\theta; \\ \dot{\theta} = g/V\left(n_y - \cos\theta\right). \end{cases} \tag{4.13}$$

By setting the restrictions control $u_{\text{lim}} = [n_{y_{\text{lim}}}]$ and solving the constraint problem for a three-dimensional state space (dependencies 4.6–4.9), we obtain the following expanded expressions—components of the constraint maintenance algorithm:

- distance to the obstacle:

$$d = \overline{a}_1 H + \overline{a}_2 L + \overline{c}; \tag{4.14}$$

- the first derivative of the range:

$$\dot{d} = V(\overline{a}_1 \sin\theta + \overline{a}_1 \cos\theta); \tag{4.15}$$

- the available acceleration of the AC braking, when the restriction system is activated:

$$\ddot{d}_a = g(n_{y_{\text{lim}}} - cos\theta)(\bar{a}_1 cos\theta - \bar{a}_2 sin\theta); \tag{4.16}$$

- correction of control at the restriction stage:

$$\Delta n_y(t) = \frac{\Delta \ddot{d}(t)}{g(\bar{a}_1 cos\theta - \bar{a}_2 sin\theta)}. \tag{4.17}$$

In accordance with the method of maintaining restrictions, the restriction algorithm is activated, when the AC approaches the obstacle by an amount:

$$d \leq d_{\text{lim},} \tag{4.18}$$

where the range of activation of the restriction system with a given control $n_{y_{\text{lim}}}$ is calculated in accordance with the expression:

$$d_{\text{lim}} = \frac{\dot{d}^2}{2\ddot{d}_a} = \frac{V^2(\bar{a}_1 sin\theta + \bar{a}_2 cos\theta)^2}{2g(n_{y_{\text{lim}}} - cos\theta)(\bar{a}_1 cos\theta - \bar{a}_2 sin\theta)}. \tag{4.19}$$

The analysis (4.19) shows a significant dependence on the activation range d_{lim} from the current speed of flight V. The larger it is, the earlier the restriction algorithm will be activated. This is also confirmed by the simulation results, presented in Fig. 4.5.

In addition, it is obvious that the ratio of the angles of inclination of the AC trajectory and the restriction line affects the range of activation of the algorithm. The position in the space of the latter is characterized by a matrix-row [2]:

$$\bar{a} = \begin{bmatrix} \frac{a_1}{|a|} & \frac{a_2}{|a|} \end{bmatrix} = [sin\mu \quad cos\mu],$$

where μ—polar angle of the line $ax + c = 0$.

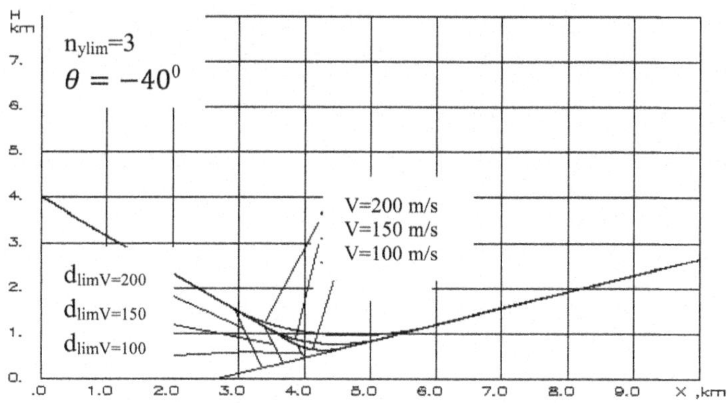

Fig. 4.5 Dependence of the activation range d_{lim} from current speed of flight V

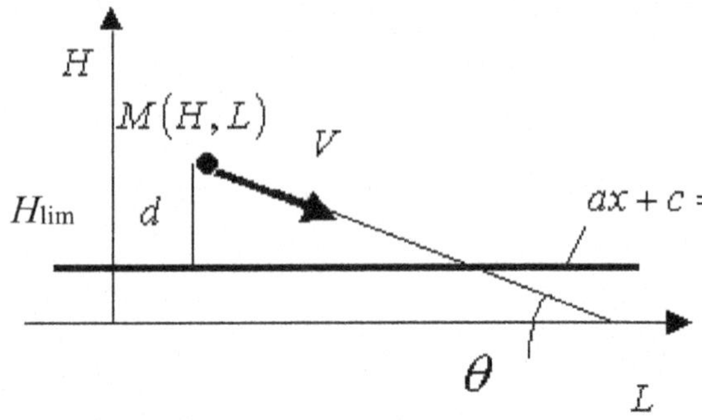

Fig. 4.6 The scheme of setting the problem of moving the AC away from the horizontal line of restriction

To show this dependence, consider the problem of moving the AC away from the horizontal line of the constraint. The scheme of the problem statement is shown in Fig. 4.6. In this case, the motion of the center of mass of the aircraft can be described by a system of equations of second order:

$$\begin{cases} \dot{H} = V sin\ \theta; \\ \dot{\theta} = g/V\left(n_{y_{\lim}} - cos\ \theta\right). \end{cases} \tag{4.20}$$

And the equation of the constraint line will be reduced to the form:

$$H - H_{\lim} = 0. \tag{4.21}$$

The expression (4.21) can be easily obtained from (4.12) by putting

$$a = [1 \quad 0]; \quad c = -H_{\lim}. \tag{4.22}$$

Substituting the values of the coefficients of the matrix-row in (4.19) and (4.17), we get the relations for the range of switching on the algorithm and the correction of control at the restriction stage:

$$d_{\lim} = \frac{(V sin\theta)^2}{2g cos\theta (n_{y_{\lim}} - cos\theta)}; \tag{4.23}$$

$$\Delta n_y(t) = \frac{\Delta \ddot{d}(t)}{g cos\theta}. \tag{4.24}$$

The analysis of the dependence (4.23) shows that the greater the initial angle of inclination of the trajectory, the earlier the constraint system will come into operation. In the limit at $\theta \to -90$ *deg*, the restriction system will be activated regardless of the initial flight altitude, preventing a vertical dive that may occur as a result of an emergency. This is also confirmed by the results of modeling the removal of the AC from a dangerous height $H_{\lim} = 200m$ (see Fig. 4.7).

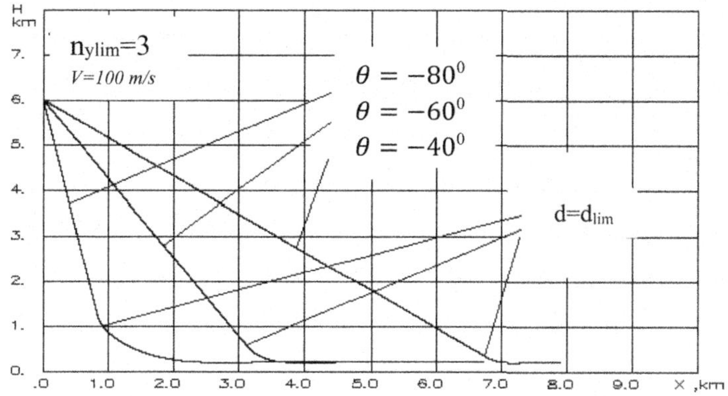

Fig. 4.7 Simulation results of the AC withdrawal from a dangerous height $H_{\lim} = 200m$

Control accounting provides a different picture $n_{x_{\lim}}$ by $V = var$.

2. The withdrawal of the aircraft from a dangerous height in the vertical plane with $V = var$.

The system of equations corresponds to this case:

$$\begin{cases} \dot{H} = V sin\,\theta; \\ \dot{\theta} = g/V\left(n_y - cos\,\theta\right); \\ \dot{V} = g(n_x - sin\,\theta). \end{cases} \tag{4.25}$$

Let's imagine the system (4.25) in the form (4.2), while $x = [H \;\; L]^T$, $w = [\theta \;\; V]^T$, and $u = [n_y\; n_x]^m$. By setting the restrictions control

$$u_{\lim} = [n_{y_{\lim}} \;\; n_{x_{\lim}}]^m \tag{4.26}$$

and using the described method of synthesis of algorithms for maintaining constraints, we have

- the range of activation of the restriction system (in this case, it corresponds to the height of activation):

$$d_{\lim} = \frac{(V sin\theta)^2}{2g(n_{x_{\lim}} sin\theta + n_{y_{\lim}} cos\theta - 1)}; \tag{4.27}$$

- correction of control at the restriction stage:

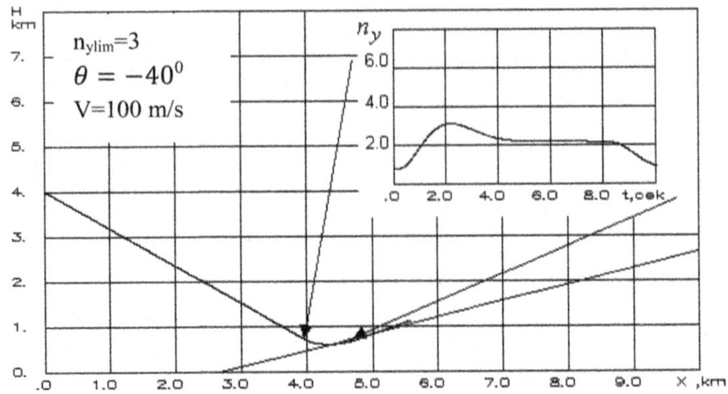

Fig. 4.8 The results of modeling the withdrawal of aircraft in the vertical plane from the inclined line

$$\Delta n_x(t) = \frac{\Delta\ddot{d}(t)}{g} sin\theta;$$

$$\Delta n_y(t) = \frac{\Delta\ddot{d}(t)}{g} cos\theta. \quad (4.28)$$

As can be seen from the resulting expressions, $\theta = -90°$ is not a special point.

It should be noted that the algorithms given here for maintaining the constraints are also effective, when studying the motion of the AC as a solid body. So, in Fig. 4.8 the results of modeling the AC withdrawal in the vertical plane from the inclined line are presented.

During the simulation, the real characteristics of the steering drives were taken into account. The callout shows the transition process of changing the normal overload on the withdrawal section. From the obtained results, it is clear that there is no error in maintaining the given limit.

Thus, the analysis of the obtained expressions and results shows evidence of the adaptive properties of the restriction algorithm to the initial conditions of the aircraft dive, which favorably distinguishes the proposed algorithms from those, used in practice.

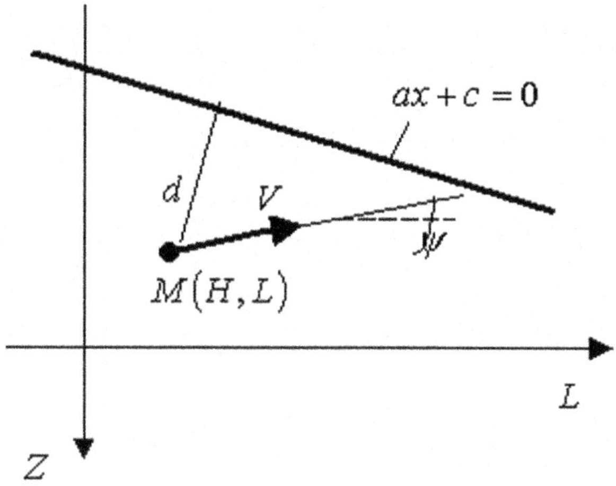

Fig. 4.9 The constraint in OLZ coordinates

4.2.2 Aircraft Drift in the Horizontal Plane

Let the motion of the AC in the horizontal plane be described by a system of equations:

$$\begin{cases} \dot{L} = V\cos\,\psi; \\ \dot{Z} = -V\sin\,\psi; \\ \dot{\psi} = -g/Vn_y \sin\,\gamma. \end{cases} \tag{4.29}$$

The constraint in OLZ coordinates is represented by a line (see Fig. 4.9):

$$a_1 Z + a_2 L + c = 0. \tag{4.30}$$

It is easy to see that the expression (4.30) is an expanded representation of the Eq. (4.12), provided that $x = [Z\ L]^m$.

The scheme of setting this problem is shown in Fig. 4.9.

To synthesize the constraint algorithms, we use the approach, described in Sect. 4.1. In accordance with it, we present the system of Eq. (4.29) in the form of (4.2), denoting the state vector $x = [Z \quad L]^T$, vector of pseudo-control $w = [\psi]$, and vector of control $u = [n_y\ \gamma]^T$. By setting the restrictions control $u_{\lim} = [n_{y\lim}\ \gamma_{\lim}]^m$ and formally using the expressions (4.6–4.9), after simple transformations, we get

- the shortest distance from the AC to the limit line:

$$d = \bar{a}_1 Z + \bar{a}_2 L + \bar{c}, \tag{4.31}$$

where $\overline{a}_1 = \frac{a_1}{|a|}$; $\overline{a}_2 = \frac{a_2}{|a|}$; $\overline{c} = \frac{c}{|a|}$;

- the speed at which the AC approaches the limit:

$$\dot{d} = V(\overline{a}_2 \cos\psi - \overline{a}_1 \sin\psi); \tag{4.32}$$

- the available acceleration of the AC braking, when the restriction system is activated:

$$\ddot{d}_a = gn_{y\,\lim} \sin_{\gamma\,\lim}(\overline{a}_1 \cos\psi + \overline{a}_2 \sin\psi); \tag{4.33}$$

- correction of control at the restriction stage:

$$\Delta u(t) = \begin{bmatrix} \Delta n_y(t) \\ \Delta\gamma(t) \end{bmatrix} =$$

$$= \left[gsin\gamma(\bar{a}_1 cos\psi + \bar{a}_2 sin\psi) \quad gn_y(\bar{a}_1 cos\psi + \bar{a}_2 sin\psi) \right]^{\#} \Delta\ddot{d}(t). \tag{4.34}$$

As a result, you can write an expression for the range of activation of the restriction system in the form:

$$d_{\lim} = \frac{\dot{d}^2}{2\ddot{d}_a} = \frac{V^2(\bar{a}_2 cos\psi - \bar{a}_1 sin\psi)^2}{2gn_{y_{\lim}} sin\gamma_{\lim}(\bar{a}_1 cos\psi + \bar{a}_2 sin\psi)}. \tag{4.35}$$

It is necessary to note the following obvious fact concerning the assignment of the constraint control. In order for the AC to perform the withdrawal maneuver strictly in the horizontal plane, the overload of withdrawal and roll must be related by a relationship:

$$n_{y_{\lim}} = 1 + \frac{1 - cos\gamma_{\lim}}{cos\gamma_{\lim}}.$$

If no special conditions are imposed on the movement of the AC in the vertical plane, when performing the withdrawal in the «horizon», then $n_{y_{\lim}}$ and $\gamma_{\lim}$ can be set for some other reasons.

The analysis (4.35) shows that the range of activation increases with increasing flight speed, and the smaller the greater the specified control limit. The results of modeling the withdrawal process in the horizontal plane according to the above algorithm are shown in Figs. 4.10, 4.11 and 4.12. The constraint line in the simulation was set by the following parameters:

$$a_1 = 1;\ a_2 = 0;\ c = 5.$$

In Fig. 4.10, the callouts show the transition process of changing the roll angle and normal overload on the withdrawal section. It shows that at the initial moment of time, the algorithm implements the specified constraint control.

Next, to maintain $\ddot{d}_a = const$, when changing the bearing angle ψ, the control is reset $\gamma_{\lim}$ with minor correction of control $n_{y_{\lim}}$.

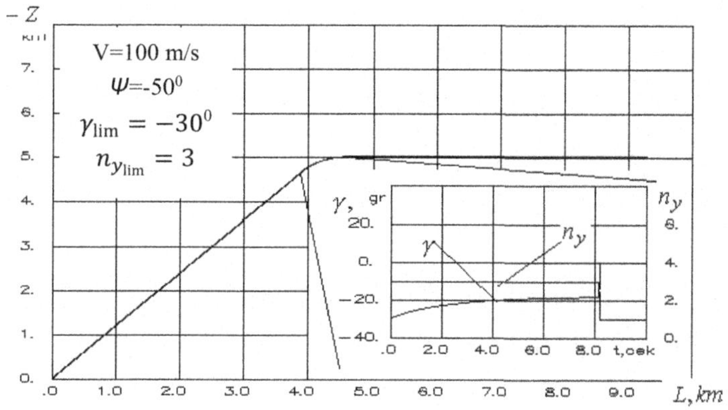

Fig. 4.10 The transition process of changing the roll angle and normal overload on the withdrawal section

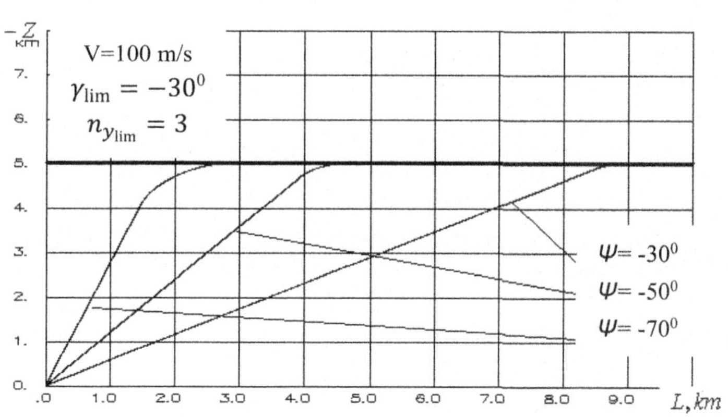

Fig. 4.11 Illustration the properties of the algorithm's adaptability to changes of the angles of approach of the aircraft to the restriction line, as well as to the control settings of the restriction -30 ° and constant value 3

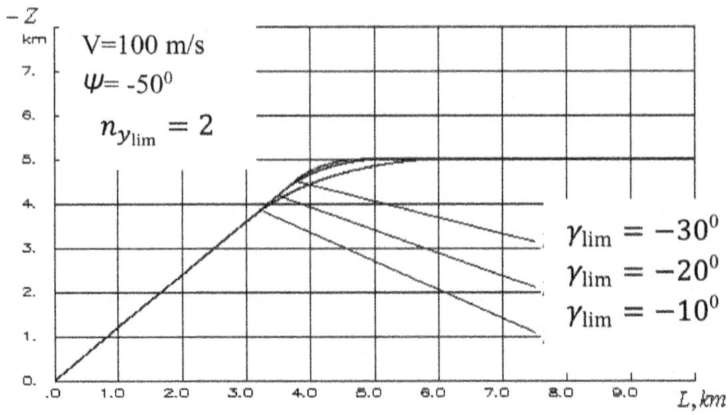

Fig. 4.12 Illustration the properties of the algorithm's adaptability to changes of the angles of approach of the aircraft to the restriction line, as well as to the control settings of the restriction -50 ° and constant value 2

Figures 4.11 and 4.12 illustrate the properties of the algorithm's adaptability to changes in the angles of the approach of the aircraft to the restriction line, as well as to the control settings of the restriction $\gamma_{\lim}$ at a constant value $n_{y_{\lim}}$.

4.3 Synthesis of Algorithms for Maintaining Limitations Within the Range of Altitudes and Flight Speeds

In the first section, it was shown that from the point of view of flight safety, the most dangerous exit is $V_{min_{\lim}}$, that is, at the left border of the range of altitudes and flight speeds. However, the synthesis of algorithms for maintaining restrictions within the range of altitudes and flight speeds causes a number of obvious difficulties, namely.

- the state space under consideration is not metric;
- one of the coordinates to be restricted, in particular the velocity, is the «control» (included on the right-hand side of the altitude equation) for the other.

The first difficulty can be solved by rationing the space. The second one introduces ambiguity in the choice of parameters for controlling the restriction.

We will deduce the main dependencies of the constraint algorithm, taking into account the above features.

Let the motion of the AC be described by the following system of differential equations:

$$\underbrace{\begin{bmatrix} \dot{H} \\ \dot{V} \\ \dot{\theta} \end{bmatrix}}_{\dot{x}} = \underbrace{\begin{bmatrix} V \sin\theta \\ g(n_x - \sin\theta) \\ g/V(n_y - \cos\theta) \end{bmatrix}}_{f(x,\, u)} \tag{4.36}$$

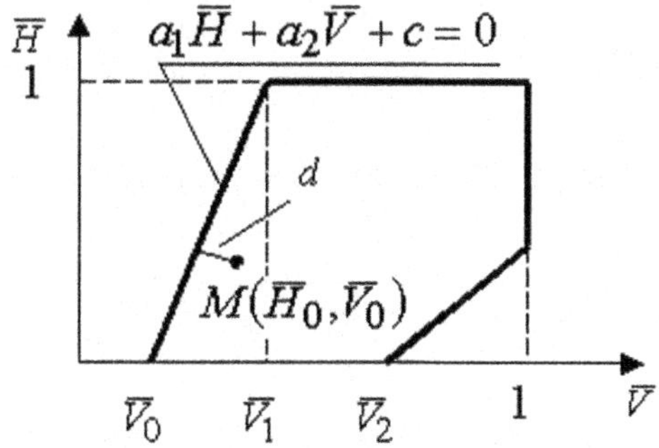

Fig. 4.13 Setting the limit of restriction

The boundary of the constraint (see Fig. 4.13) is set in a slice of space X for relative coordinates ($\bar{H}$, $\bar{V}$). And under $\bar{H}$ and $\bar{V}$, we will understand

$$\bar{H}=\frac{H}{\bar{H}_{max}\frac{V}{V_{max}}}.$$

Consider the case of approaching a point $M(\overline{H}_0,\ \overline{V}_0)$, describing the position of the AC in the considered coordinate system, to the left border of the range of altitudes and flight speeds.

The equation of the latter can be represented by an expanded expression:

$$a_1\bar{H}+a_2\bar{V}+c=0, \tag{4.37}$$

where according to Fig. 4.13: $a_1 = \overline{V}_1 - \overline{V}_0$; $a_2 = -1$; $c = \overline{V}_0$.

As stated earlier, the distance to the range boundary (4.37) is defined as

$$d = \bar{a}_1\overline{H} + \bar{a}_2\overline{V} + \bar{c}. \tag{4.38}$$

In (4.38),

$$\bar{a}_1=\frac{a_1}{\sqrt{{a_1}^2+{a_2}^2}};\quad \bar{a}_2=\frac{a_2}{\sqrt{{a_1}^2+{a_2}^2}};\quad \bar{c}=\frac{c}{\sqrt{{a_1}^2+{a_2}^2}}.$$

Let's take the first and second time derivatives t from the distance d and analyze the resulting expressions. So, taking into account that

$$\dot{d} = \dot{d}_H + \dot{d}_V;\ \ddot{d}_a = \ddot{d}_{aH} + \ddot{d}_{aV},$$

for the component of the distance change in height, we have

$$\dot{d}_H=\frac{\bar{a}_1}{H_{max}sin\theta}; \tag{4.39}$$

$$\ddot{d}_H = \frac{\bar{a}_1 g}{H_{max\,x} sin\theta_y cos\theta}. \tag{4.40}$$

Accordingly, the first and second derivatives of the velocity component are equal:

$$\dot{d}_V = \frac{\bar{a}_2 g}{V_{max\,x} sin\theta}; \tag{4.41}$$

$$\ddot{d}_V = \frac{\bar{a}_2 g}{V_{max}\left(\dot{n}_x - \frac{g}{V} cos\theta\left(n_y - cos\theta\right)\right)}. \tag{4.42}$$

Analysis of the obtained expressions, and in particular (4.40), (4.42), shows that the general control, affecting the acceleration of braking $\ddot{d}_{aH}$, $\ddot{d}_{aV}$, is a normal overload n_y and the derivative of the longitudinal overload $\dot{n}_x$. To use the control $\dot{n}_x$ correctly, the system (4.36) must be extended by the equation:

$$\dot{n}_x = \dot{n}_{x\lim.}$$

Thus, for the object, described by the extended system of Eqs. (4.36), the control of the constraint will be given by the vector:

$$u_{\lim} = [\dot{n}_{x\lim}\ n_{y\lim}]^m. \tag{4.43}$$

Then the distance to the boundary of the constraint, at which the constraint algorithm must be turned on in the operation, and the compensation control is determined by the following expressions:

$$d_{\lim} = \frac{\dot{d}^2}{2\ddot{d}_a} = \frac{(\dot{d}_V + \dot{d}_H)^2}{2[\ddot{d}_V(\dot{n}_{x\lim}, n_{y\lim}) + \ddot{d}_H(n_{y\lim})]}; \tag{4.44}$$

$$\Delta u(t) = \left[\frac{\overline{a}_2 g}{V_{max} \cos\theta \left(\frac{\overline{a}_1}{H_{\max} \frac{\overline{a}_2 g}{V V_{\max}}}\right)^{\#}}\right]. \tag{4.45}$$

In Fig. 4.14, the results of modeling the restriction process within the range of altitudes and flight speeds, when the aircraft enters the left boundary of the range, described by the expression (4.38) are presented.

Initial flight mode: $H = 2$км, $V = 1300$км/h. We studied the climb of altitude with different angles of inclination of the trajectory and the loss of speed. There are adaptive properties of the algorithm to the initial conditions. Thus, an increase in the angle of inclination of the trajectory leads to an earlier activation of the constraint algorithm.

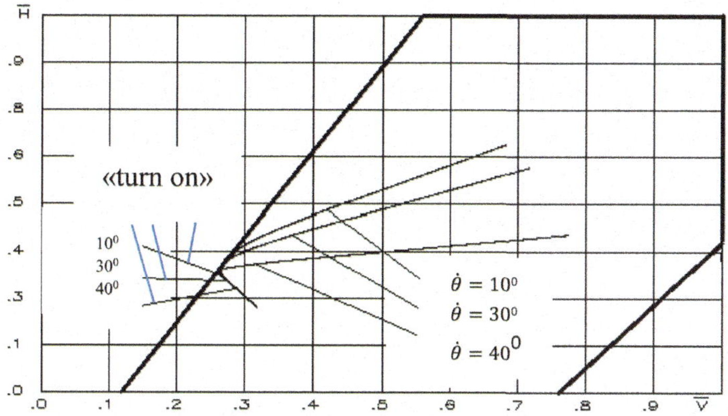

Fig. 4.14 Simulation results of the restriction process within the range of altitudes and flight speeds, when the aircraft enters the left border of the range

4.4 Features of Maintaining Non-Stationary Constraints

In the study of trajectory problems and in real practice, it is often necessary to limit the flight parameters, depending on the presence of moving objects of different natures. These are, for example, AC, flying on intersecting courses, areas of thunderstorm activity, and so forth. A number of papers have been devoted to solving this problem [3–7]. However, the task has not been fully solved. In this section, using a fairly simple, but not trivial, example, the applicability of the described method to solving the problem of limiting the non-stationary boundary is shown.

Let two material points, and one of them is controlled, move toward each other. The equations, describing their motion, are presented in the following form:

- equations of motion of the controlled object of the restriction:

$$\begin{cases} \dot{X} = V; \\ \dot{V} = a = const; \end{cases} \tag{4.45}$$

- equations of motion of the border:

$$\begin{cases} \dot{X}_{\text{lim}} = V_{\text{lim}}; \\ \dot{V}_{\text{lim}} = a_{\text{lim}}; \\ \dot{a}_{\text{lim}} = const. \end{cases} \tag{4.46}$$

You need to determine the distance between the object and the moving border, when reaching which you need to turn on the specified control of restrictions for the object (4.45). All parameters of the motion of the non-stationary boundary are considered known.

According to the method of restriction, the distance to the border is defined as

$$d = X_{\text{lim}} - X; \tag{4.47}$$

the first derivative of the distance:

$$\dot{d} = \dot{X}_{\text{lim}} - \dot{X} = V_{\text{lim}} - V; \tag{4.48}$$

acceleration of braking, provided that the control of restriction is turned on a_{lim}:

$$\ddot{d}_a = \ddot{X}_{\text{lim}} - \ddot{X}(a_{\text{lim}}); \tag{4.49}$$

$$\Delta a(t) = -\Delta\ddot{d}(t) = -\left(a_{\text{lim}} - a_{\text{lim}_{\text{mean}}}\right), \tag{4.50}$$

where $a_{\text{lim}_{\text{mean}}}$—acceleration of the boundary, when the constraint algorithm is turned on.

As a result, the desired distance of the turning on of the restriction algorithm is in accordance with the expression:

$$d_{\text{lim}} = \frac{\dot{d}^2}{2\ddot{d}_a} = \frac{(V_{lim} - V)^2}{2(a - a_{\text{lim}})}. \tag{4.51}$$

The simulation of the constraint process, using the above algorithms, is illustrated in Fig. 4.15.

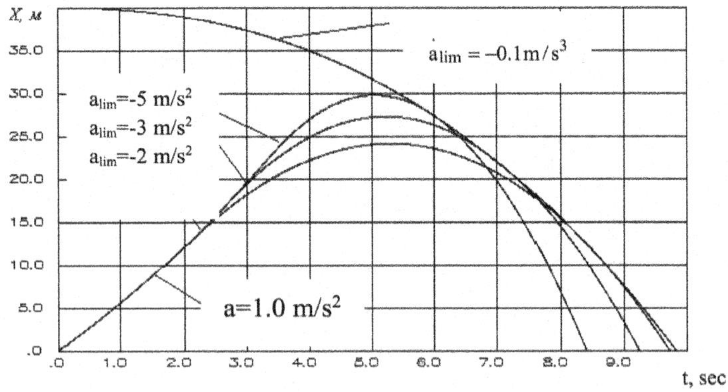

Fig. 4.15 The simulation of the constraint process, using the above algorithms

From the above results, it can be seen that the constraint algorithm is also workable, when considering the non-stationary boundary. The larger the control of the restriction, the faster the object can approach it. However, to ensure the restriction process, you need to know the parameters of the movement of the border itself.

This approach can also be applied to more complex cases of describing both the constraint boundary and the control object.

References

1. Vygodsky M Ya (1958) Handbook of higher mathematics. - M.: State publishing house of technical and theoretical literature
2. Gantmacher FR (1988) Matrix theory. - Moscow: Nauka
3. Akimov AN, Andreev VV (1997) Method of direct optimization. The main provisions and the order of application. - M: Air Force Engineering Academy named by N.E. Zhukovsky
4. Akimov AN, Vorobyov VV (2001) Methods and algorithms of aircraft withdrawal from the spatial surface of the restriction. Automation and telemechanics, No. 7
5. Akimov AN, Vorobyov VV, Lukashov SV (1998) Algorithm for preventing an airplane from colliding with the Earth's surface. Problems of flight safety. - M.: VINITI, p 47–50
6. Akimov AN, Vorobyov VV, Lukashov SV (1999) Prevention of a collision of an aircraft with the terrain by leading away along spatial trajectories. Proceedings of Scientific readings, dedicated to the memory of aircraft designer I. I. Sikorsky. Moscow – St. Petersburg, May 25–29, 3–1
7. Akimov AN, Vorobyov VV, Fedyunin M Yu (1995) Algorithm of functioning of the adaptive system of limiting the angle of attack. Ensuring flight safety: Scientific and methodological materials. – M.: Air Force Engineering Academy named by NE Zhukovsky

Chapter 5
Aircraft Drift Away from Limiting Surfaces Along Programmed Trajectories

5.1 Substantial Problem Statement

This section presents an approach to the construction of algorithms for systems for maintaining restrictions in the trajectory circuit of aircraft's control, based on the preliminary assignment of the trajectory of the AC movement.

The use of this approach was a consequence of the specific actions of a person as an operator in stressful situations, and it is known that the crew's activities in flight at low altitudes belong to the first type of extreme situations.

Taking into account the psychology of the pilot, it is obvious that the automatic withdrawal of the AC from the earth's surface should be performed only by performing ascending maneuvers (it is also possible to maneuver the withdrawal strictly in the horizontal plane, but not descending).

In Fig. 5.1, an example of the constraint problem is shown schematically, where the algorithms, considered above, will try to make a withdrawal from the boundary by a descending maneuver—line 1 in the figure, as opposed to the desired one, line 2.

To solve this problem, let's consider the scheme of AC withdrawal in the normal coordinate system $OX_gY_gZ_g$ from the constraint surface, given by equation [1–3]:

$$\xi(x,\ y,\ z) = 0 \tag{5.1}$$

and introduce the following notation (see Fig. 5.2).

$M(X_0, Y_0, Z_0)$—point with current coordinates of AC; $T(X, Y, Z)$—the point on the surface of the constraint, obtained by the intersection of the continuation line of the velocity vector of the AC with the surface; K—a plane, tangent to the constraint surface at a point T; U—some plane, in which the obstacle evasion maneuver is performed, containing the normal $\vec{N}_k$ to the surface in a point T; B and G—vertical and horizontal planes, respectively; S—the line of intersection of the withdrawal plane U with the plane K; ζ—the angle between the plane U and the vertical plane B; η—the angle between the straight line S and the horizontal plane G.

V. Akimov and V. A. Zatuchny, *Limiting Modes of Aircraft Flight*,
Springer Aerospace Technology, https://doi.org/10.1007/978-981-19-6329-2_5

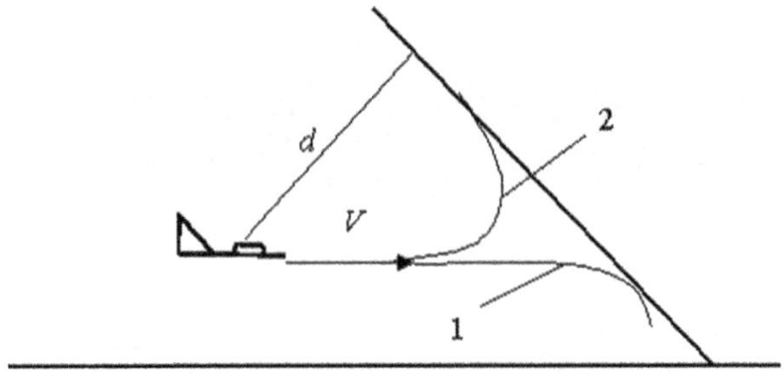

Fig. 5.1 Example of the restriction task

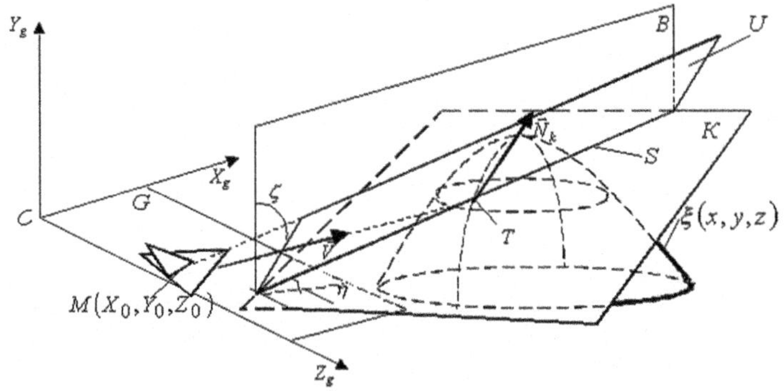

Fig. 5.2 The scheme of withdrawal of the aircraft in the normal coordinate system $OX_gY_gZ_g$ from the constraint surface

The control of the AC in the process of withdrawal is carried out in the channels of roll and pitch in order to bring the object into the plane of withdrawal U and turn it in it with the maximum available capabilities, until the AC flight trajectory is aligned with the S line. At the same time, it is assumed that the command control signals are generated by the AC collision warning system with the earth's surface.

5.2 Algorithms for Calculating the Angles of the Aircraft Position in Space in the Case Movement It Away from an Obstacle

It is obvious that the angular position of the AC in space during the withdrawal is directly dependent on the orientation of the planes U and K, defined, in particular, by the angles ζ and η (see Fig. 5.2), which in turn depend on the type of the constraint surface and the position of the point T on it.

The coordinates of the point T are determined by jointly solving the equations of the surface (5.1) and the line, containing the velocity vector AC. The canonical representation of the latter has the form:

$$\frac{x - X_0}{V_x} = \frac{y - Y_0}{V_y} = \frac{z - Z_0}{V_z}. \tag{5.2}$$

Here, V_x, V_y, V_z—projections of velocity vector $\overrightarrow{V}$ on the axis of the coordinate system under consideration.

Then the plane, tangent to the constraint surface at point T, can be represented by the equation:

$$\xi_x^{'}(x - X) + \xi_y^{'}(y - Y) + \xi_z^{'}(z - Z) = 0, \tag{5.3}$$

where $\xi_x^{'} = \frac{\partial \xi(x,y,z)}{\partial x}$, $\xi_y^{'} = \frac{\partial \xi(x,y,z)}{\partial y}$, $\xi_z^{'} = \frac{\partial \xi(x,y,z)}{\partial z}$—the values of the partial derivatives of the constraint surface at the point $T\ (X, Y, Z)$ according to the corresponding parameters.

In the case of considering a concave surface as a constraint surface (here, a concave surface is understood as a surface, whose all points relative to the point T of the tangent surface in some given neighborhood are at a smaller distance from the AC, than the point T). For physical reasons, the K plane must be constructed, using information about the coordinates of the three points, closest to AC on the constraint surface. These points can be obtained, using the onboard means of the AC. Then the equation of the plane K will have the form:

$$\begin{vmatrix} x - X_1 & y - Y_1 & z - Z_1 \\ X_2 - X_1 & Y_2 - Y_1 & Z_2 - Z_1 \\ X_3 - X_1 & Y_3 - Y_1 & Z_3 - Z_1 \end{vmatrix} = 0. \tag{5.4}$$

For the convenience of further transformations, we write the Eqs. (5.3) and (5.4) in the form:

$$A_K x + B_K y + C_K z + D_K = 0, \tag{5.5}$$

where for Eq. (5.3):

$$A_K = \xi_x^{'},\ B_K = \xi_y^{'},\ C_K = \xi_z^{'},$$

and in the case of using Eq. (5.4):

$$A_K = \begin{vmatrix} Y_2 - Y_1 & Z_2 - Z_1 \\ Y_3 - Y_1 & Z_3 - Z_1 \end{vmatrix},$$

$$B_K = -\begin{vmatrix} X_2 - X_1 & Z_2 - Z_1 \\ X_3 - X_1 & Z_3 - Z_1 \end{vmatrix},$$

$$C_K = \begin{vmatrix} X_2 - X_1 & Y_2 - Y_1 \\ X_3 - X_1 & Y_3 - Y_1 \end{vmatrix},$$

and for both cases:

$$D_K = -(XA_K + YB_K + ZC_K).$$

When the plane U is oriented in space, it is taken into account that the speed of the AC moving away from the obstacle should have the maximum value. This can be achieved in the case, where the plane U contains the normal $\vec{N}_k$ to the plane K at the point T and the point with the current coordinates AC $M(X_0, Y_0, Z_0)$.

Then the parameters of the equation:

$$A_U x + B_U y + C_U z + D_U = 0, \quad (5.6)$$

describing the position of the U plane, calculated from the dependencies:

$$A_U = C_K(Y - Y_0) - B_K(Z - Z_0);$$

$$B_U = A_K(Z - Z_0) - C_K(X - X_0);$$

$$C_U = B_K(X - X_0) - A_K(Y - Y_0);$$

$$D_U = -(A_U X_0 + B_U Y_0 + C_U Z_0).$$

At the same time, the angle ς between the withdrawal plane U and vertical plane B (see Fig. 5.2) is equal:

$$\varsigma = arccos\left(\frac{A_B A_K + B_B B_K + C_B C_K}{\sqrt{A_B{}^2 + B_B{}^2 + C_B{}^2}\sqrt{A_K{}^2 + B_K{}^2 + C_K{}^2}}\right), \quad (5.7)$$

Where A_B, B_B, C_B—coefficients, describing the orientation of the plane B in space.

Taking into account that by definition $A_B = 1, C_B = B_B = 0$, the expression (5.7) is written as

$$\varsigma = arccos\left(\frac{A_K}{\sqrt{A_K{}^2 + B_K{}^2 + C_K{}^2}}\right). \quad (5.8)$$

In order for the AC not to enter the spiral and be held in the U plane, the forces, perpendicular to this plane (see Fig. 5.3), must be balanced.

This is provided by turning the AC to the angle of the roll-off [4]:

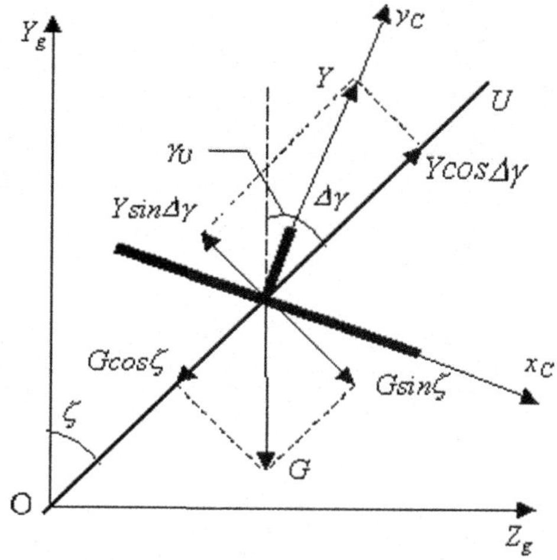

Fig. 5.3 Illustration of the force balance

$$\Delta\gamma = arcsin\frac{sin\zeta}{n_y}. \tag{5.9}$$

As a result, the required roll of the aircraft will be equal to

$$\gamma_U = \zeta - \Delta\gamma sign(\zeta). \tag{5.10}$$

The angle of inclination of the trajectory of the AC during the withdrawal is the angle η (see Fig. 5.2) between straight line *S* and plane *G*, that is, $\theta_S = \eta$.

Since the line S is formed by the intersection of the planes *U* and *K* and passes through the point *T*, then, according to [5], its canonical representation has the form:

$$\frac{x - X}{l_S} = \frac{y - Y}{m_S} = \frac{z - Z}{n_S}, \tag{5.11}$$

where $\boldsymbol{l_S},\ \boldsymbol{m_S},\ \boldsymbol{n_S}$ are determined by the relations:

$$l_S = B_U C_K - C_U B_K;\ m_S = C_U A_K - A_U C_K;\ n_S = A_U B_K - B_U A_K.$$

As a result, value θ_S is calculated as the angle between the straight line and plane, using formula

$$\theta_S = arcsin\left(\frac{m_S}{\sqrt{l_S^2 + m_S^2 + n_S^2}}\right). \tag{5.12}$$

As a result, expressions are obtained for determining and current recalculation of the specified angles of the AC position in space during the maneuver of evading from the constraint surface.

5.3 Conditions for the Activation of «Withdrawal» Mode

The system is activated, when the following conditions are met (see Fig. 5.4):

$$D_{cur.} = D_{act}. \tag{5.13}$$

where D_{act}—instantaneous radius of rotation of the center of mass AC in the plane of the withdrawal maneuver on conditions that it is performed; $D_{cur.}$—current distance from the center $C(X_C, Y_C, Z_C)$ of the curvature of the AC trajectory, when it is moved away with the specified control, up to the K plane (in Fig. 5.4 изображен the trace of the plane K, obtained by the intersection of the plane K and the plane U—line S is depicted).

The instantaneous radius of rotation of the aircraft in the plane of withdrawal, according to [4], is determined from the expression:

$$D_{\text{act.}} = \frac{V^2}{g\sqrt{n_y^2 - 2n_y cos\gamma_U cos\theta_0 + cos^2\theta_0}}, \tag{5.14}$$

where $n_y = n_{y_{\text{withdrawal}}}$.

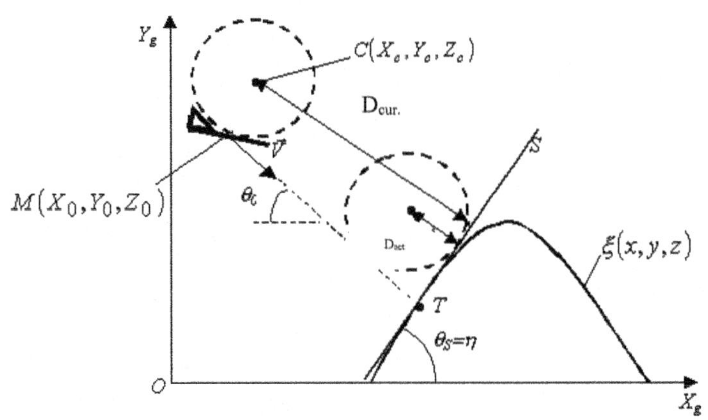

Fig. 5.4 The trace of the plane K, obtained by the intersection of the plane K and the plane U—line S

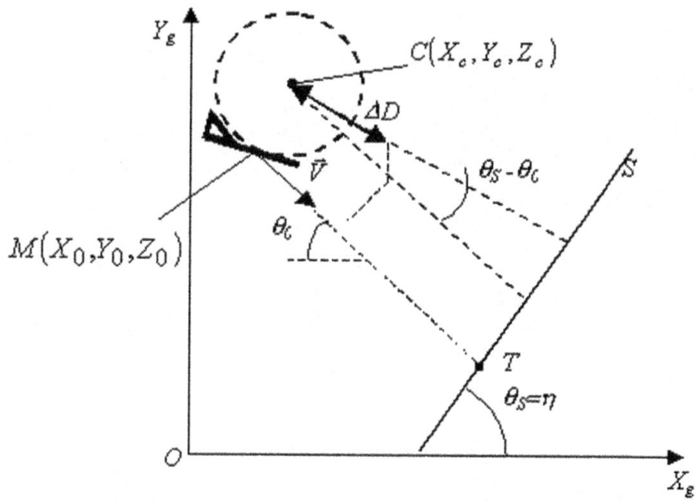

Fig. 5.5 Determination of the distance, traveled by the center of rotation C in the process of reaching the AC of the necessary overload of the withdrawal $n_{y_{\text{withdrawal}}}$ and given roll γ_U

Due to the fact that the value of distance, determined by (5.14), does not take into account the time to create an overload of withdrawal $n_{y_{\text{withdrawal}}}$ and roll γ_U, the distance of activation is corrected as follows:

$$D^*_{\text{act.}} = D_{\text{act.}} + \Delta D, \tag{5.15}$$

where $\Delta D \approx V t_{\text{cr}} sin(\theta_S - \theta_0)$—distance, traveled by the center of rotation C (see Fig. 5.5) in the process of reaching the AC of the required overload of withdrawal $n_{y_{\text{withdrawal}}}$ and given roll γ_U; $t_{\text{cr}} = t_{n_y} + t_{\gamma_U}$—total time of their creation.

Available values for the current flight mode t_{n_y}, t_{γ_U}, and $n_{y_{\text{withdrawal}}}$ can be obtained using predictive models of the short-period motion of the AC [6], feeding to their inputs the speed and deviation of the steering surfaces, trusted to the system.

Value of the current distance D_{cur} is found, using the ratio:

$$D_{\text{cur}} = \frac{A_K X_C + B_K Y_C + C_K Z_C + D_K}{\sqrt{A_K{}^2 + B_K{}^2 + C_K{}^2}}, \tag{5.16}$$

where coordinates X_C, Y_C, and Z_C are determined by the center of curvature of the spatial line [5] and have the following expressions:

$$X_C = X_0 + \frac{V_x{}^2 + V_y{}^2 + V_z{}^2}{a^2 + b^2 + c^2}\left(bV_z - cV_y\right);$$

$$Y_C = Y_0 + \frac{V_x{}^2 + V_y{}^2 + V_z{}^2}{a^2 + b^2 + c^2}(cV_x - aV_z);$$

$$Z_C = Z_0 + \frac{{V_x}^2 + {V_y}^2 + {V_z}^2}{a^2 + b^2 + c^2}(aV_y - bV_x).$$

Here, $a = V_y\dot{V}_z - V_z\dot{V}_y$, $b = V_z\dot{V}_x - V_x\dot{V}_z$, $c = V_x\dot{V}_y - V_y\dot{V}_x$; $\dot{V}_x, \dot{V}_y, \dot{V}_z$—projections of the acceleration vector of the AC on the axis of the normal coordinate system (calculated values, if there is a control signal from the «withdrawal» system).

The book uses the basic principles and algorithms of the method of direct optimization, described in [7], to implement the trajectory of withdrawal.

5.4 Formation of Controls

Consider a dynamic object of the form:

$$\dot{x} = F(x, u, t), \tag{5.17}$$

Where $x = [\theta\gamma\ n_y\omega_x\omega_y\omega_z\phi_{st}\phi_{dif}]^T$—state vector; u—vector of controls.

Let's imagine the equations, included in the system (5.7), in the form of a three-level hierarchy of system of equations:

$$\dot{z} = f(z, y, t);$$

$$\dot{y} = \psi(y, \delta, t);$$

$$\dot{\delta} = \varphi(\delta, u, t). \tag{5.18}$$

The first level of the hierarchy is represented by the specified trajectory parameters of the AC, which are included in the state vector $z = [\theta\ \gamma]^T$, and by the vector of pseudo-control $y = \left[n_y\omega_x\omega_y\omega_z\right]^T$. The second level is the flight parameters of the AC, which are included in the state vector y, and the pseudo-control vector $\delta = [\phi_{\mathrm{st}}\,\phi_{\mathrm{dif}}]^T$. The third level is a system for preventing the AC from colliding with the ground surface in terms of the deviation of the steering surfaces, taking into account the real properties of the AC steering drives. Here, δ—state vector of the system; u—the required vector of controls.

In accordance with the main provisions of the method of direct optimization, the determination of the pseudo-controls $\Delta y_i, \Delta\delta_i$, which are necessary for the withdrawal of AC from where the surface of the increment's constraint is carried out, using the relations:

$$\Delta y_i = \left(\left.\frac{\partial f}{\partial y^T}\right|_i\right)^{\#}(\dot{z}_* - f_i(z, y, t));$$

$$\Delta\delta_i = \left(\frac{\partial\psi}{\partial\,\delta^T}\bigg|_i\right)^{\#}(\dot{y}_* - \psi_i(y,\delta,t));$$

$$\Delta u_i = \left(\frac{\partial\varphi}{\partial\,u^T}\bigg|_i\right)^{\#}\big(\dot{\delta}_* - \varphi_i(\delta,u,t)\big),$$

where #—pseudo-circulation sign;$\dot{z}_* = \frac{z_k - z(t_i)}{h_z}$,$\dot{y}_* = \frac{\Delta y_i}{h_y}$, $\dot{\delta}_* = \frac{\Delta\delta_i}{h_\delta}$—required values of the velocity vectors, determined at each i-th step from the condition that the system reaches the state $z_k = \left[\theta_S\,\gamma_U\right]^{\mathrm{T}}$; h_z, h_y, h_δ—the control intensities, which are set, using the acceptable quality of the transition processes;$\frac{\partial f}{\partial y^T}$,$\frac{\partial\psi}{\partial\delta^T}$, $\frac{\partial\varphi}{\partial u^T}$—the matrices of partial derivative, which are recalculated at each i-th step of forming the controls.

5.5 Solving Private Problems of Aircraft Evacuation from Obstacles

This section presents the results of modeling the movement of an aircraft in conditions of a possible collision with various types of restriction surfaces. We used the initial information about its typical aerodynamic, dynamic, geometric, and mass characteristics, described in [6].

5.5.1 Vertical Wall

Let the AC move from the starting point of space M (Fig. 5.6) with coordinates (X_0, Y_0, Z_0). In this case, the parameters of its movement are known: speed of flight V_0, the angle of incline of the trajectory θ_0, roll γ_0, and maximum available overload $n_y^{withdrawal}$.

It is required, for a given known position of the obstacle, to determine the distance of activation, starting from which the withdrawal system would be able to prevent a collision with the «wall» with overload $n_y^{withdrawal}$.

In this case, the constraint surface is represented by an equation of the first degree:

$$Ax + By + Cz + D = 0,$$

where $A = 1$, $B = 0$, $C = 0$, and $D = -3$.

The current values of the roll and the angles of the trajectory's incline are taken as the initial conditions:

$$\gamma_0 = \gamma(t_{cur}),\ \theta_0 = \theta(t_{cur}).$$

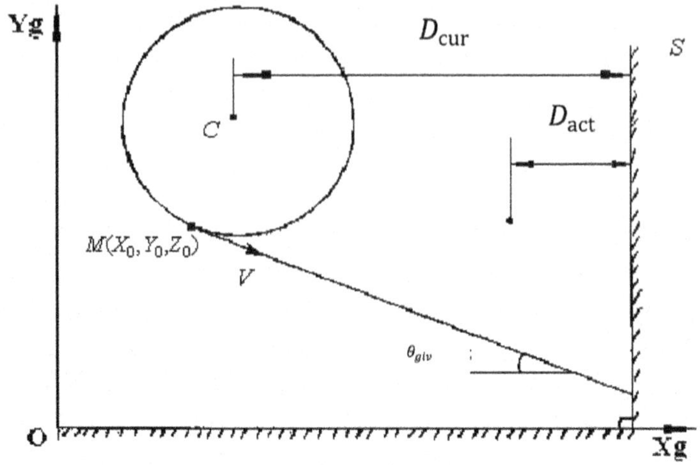

Fig. 5.6 Model of the motion of the AC from the starting point of space with known motion parameters

Their specified final values depend on the course angle φ of the approach to the wall (see Fig. 5.7). By it, we mean the angle between the projection of the velocity vector of the AC on the horizontal plane $X_g O_g Z_g$ and straight line of intersections of the wall with the horizon L.

Obviously, in the case of $\phi = 0\,deg$ the flight is carried out parallel to the obstacle on the left or right. Therefore, due to the absence of a point T on the constraint

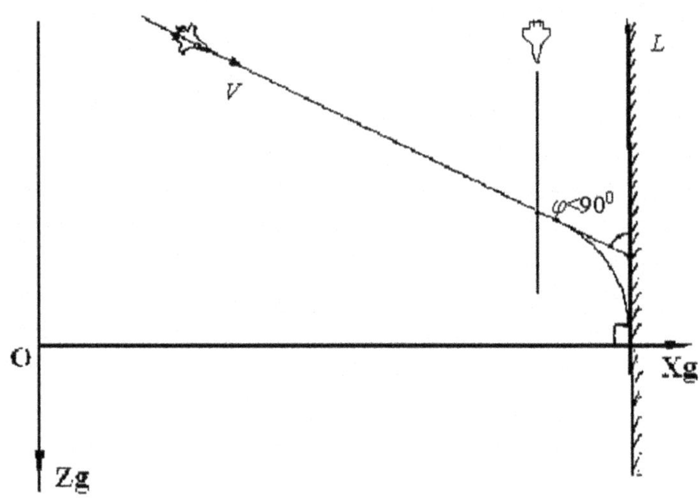

Fig. 5.7 Dependence of the final values of the angles of roll and incline of the trajectory on the course angle φ of approach to the wall

surface, obtained by the intersection of the continuation line of the velocity vector AC with the surface, the algorithm will never enter into operation.

When $\phi = 90\ deg$, the object's flight trajectory runs along the normal to the obstacle. The final conditions, according to dependencies (5.10) and (5.12), take the following values:

$\gamma_{giv} = 0\ deg, \theta_{giv} = 90\ deg.$

Figures 5.8 and 5.9 show the results of the conducted studies on the evaluation of the algorithm's performance, when $\phi = 90\ deg$. The initial values of the dive angle and the number of flight M were varied.

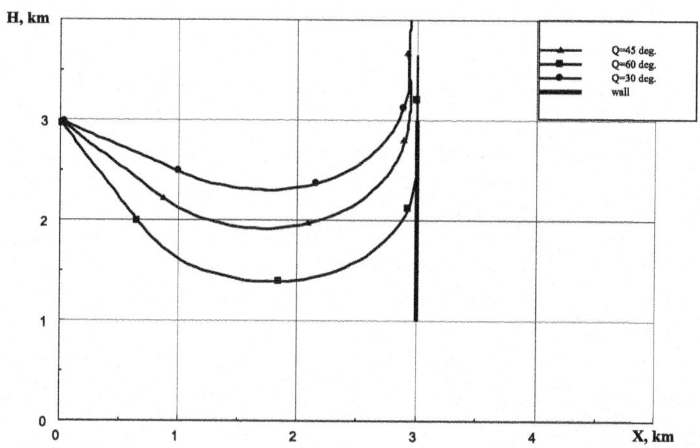

Fig. 5.8 The results of the conducted research on the evaluation of the algorithm performance at $\phi = 90\ deg$ (M = 0,7)

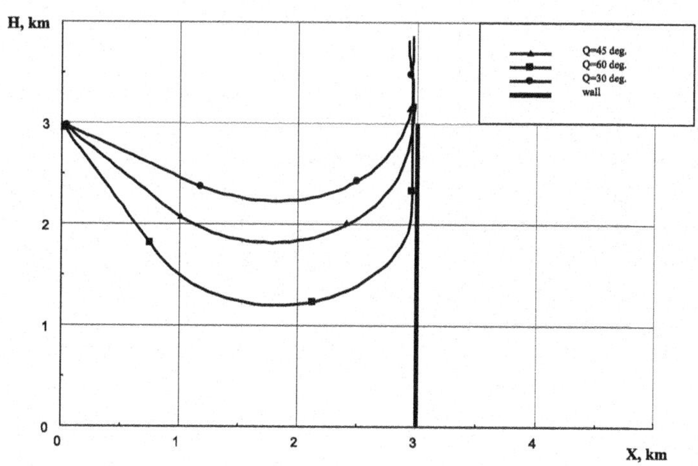

Fig. 5.9 The results of the conducted research on the evaluation of the algorithm performance at $\phi = 90\ deg$ (M = 0,8)

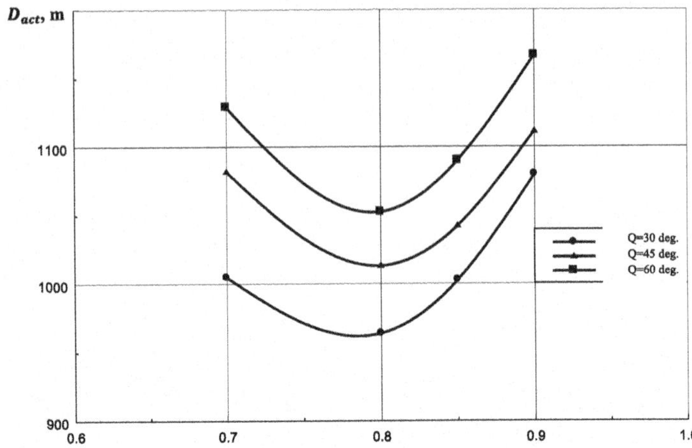

Fig. 5.10 Changing value D_{act} at different values of the flight's speed

It is obvious that changing the initial conditions in a sufficiently wide range does not lead to the piercing of the object on the constraint surface.

Changing of value D_{act} at different values of the flight speed has a character close to the quadratic one (see Fig. 5.10). The presence of the minimum function for the number of flight M near the value 0.8 can be explained by the possibility of creating at a given flight speed the maximum available overload of withdrawal, equal to the maximum permissible overload for the AC. Therefore, at M > 0.8, an increase in the flight speed has a decisive influence on the value of the distance of activation.

5.5.2 *Inclined Plane*

The construction of the maneuver for moving the AC away from the inclined plane K is strongly dependent on the course angle φ of the approach to the obstacle. So, when $\phi = 90\ deg$ (see Fig. 5.11), it is obvious that $\gamma_{giv} = 0\ deg$ and $\theta_{giv} = \eta$.

When the object is flying along a straight line r of the intersection of the planes K and G (see Fig. 5.12), the given roll and angle of inclination of the trajectory take the following values: $\theta_{giv} = 0$; $\gamma_{giv} = \eta - \Delta\gamma \cdot sign(\eta)$.

The general case of constructing the given angles for the withdrawal maneuver is shown in Fig. 5.13.

The results of studies on the withdrawal of the AC from the inclined plane with different initial values of the dive angle θ_0 and the number M of flight, as well as variations in the inclination of the restriction plane to the horizon, are shown in Figs. 5.14 and 5.15.

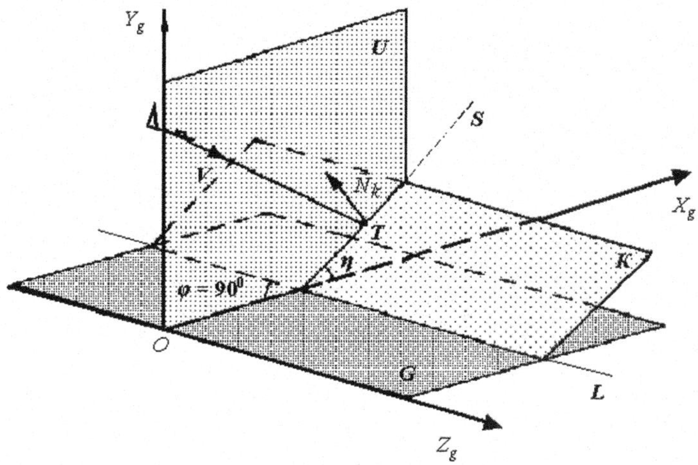

Fig. 5.11 Construction of the maneuver of moving the AC away from the inclined plane with the course angle $\phi = 90\ deg$ of approach to the obstacle

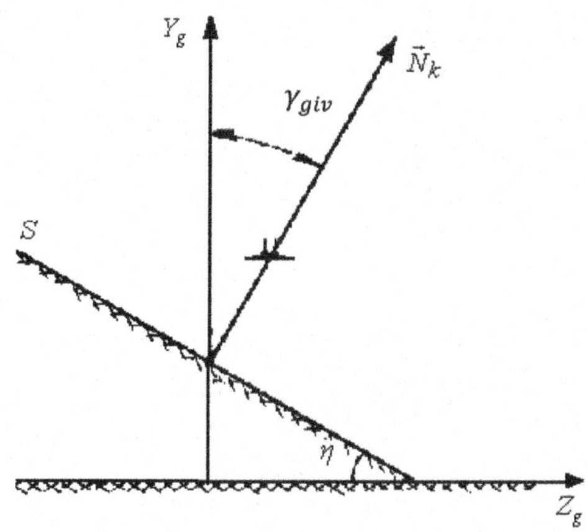

Fig. 5.12 Flight along a straight line *r* of the intersection of the planes *K* and *G*

It can be seen that changing the flight conditions and the orientation parameters of the restriction plane in space does not lead to a violation of the performance of the developed algorithm.

5.5.3 Comparative Analysis of Algorithm «Withdrawal»

The comparative analysis of the proposed algorithm with the algorithms currently used as traditional «Withdrawal» modes was carried out without taking into

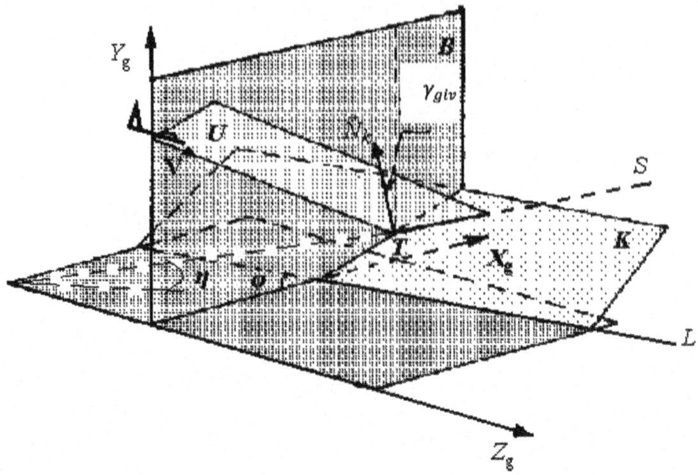

Fig. 5.13 The general case of constructing the given angles for the withdrawal maneuver

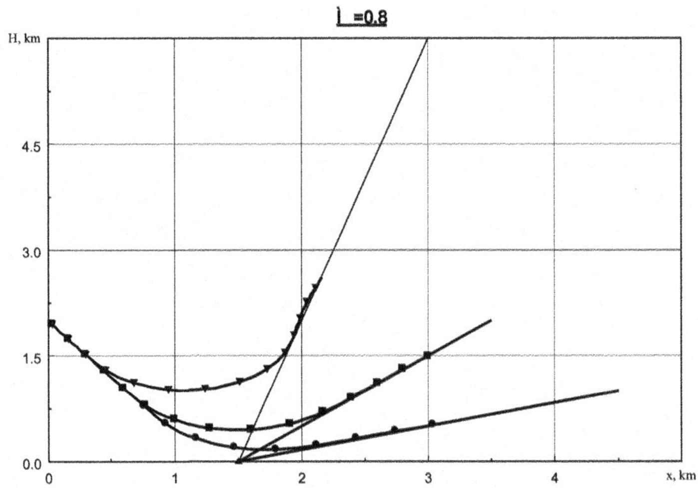

Fig. 5.14 Results of studies on the withdrawal of the AC from the inclined plane (M = 0,8)

account modeling errors, atmospheric turbulence, and errors of onboard measuring equipment.

The task was to prevent the collision of an object with an inclined plane of restriction, «located» at a distance of 1500 m from the beginning of the earth's coordinate system, by moving it away in the vertical plane.

In Figs. 5.16, 5.17, 5.18 and 5.19, the trajectories of the object movement during the maneuver of moving away from the inclined planes of the constraint with different angles of inclination to the horizon η are presented.

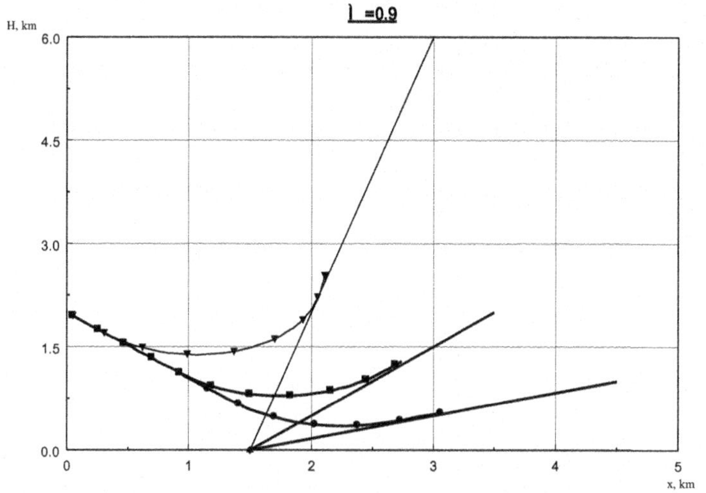

Fig. 5.15 Results of studies on the withdrawal of the AC from the inclined plane (M = 0,9)

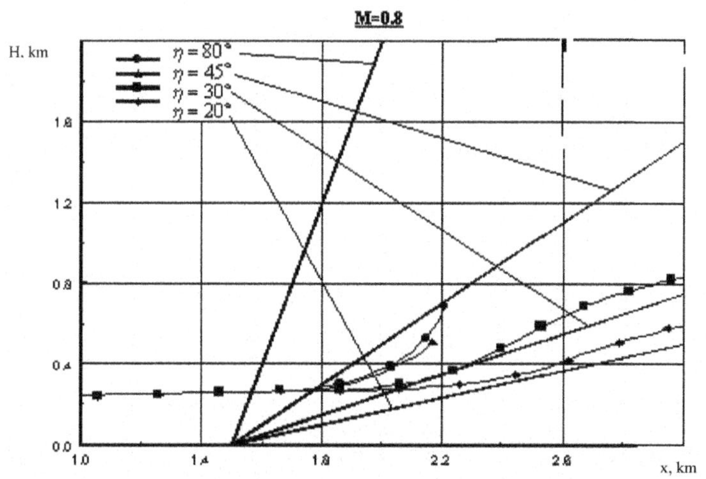

Fig. 5.16 Trajectories of the object movement during the maneuver of moving away from the inclined planes of restriction

Studies show that traditional algorithms are not sufficiently efficient (see Figs. 5.16 and 5.17), when a withdrawal of the AC from the inclined plane. So, the «piercing» of an obstacle by an object occurs already, when the angle η is greater than 18 degrees. The main reason is that there is no correction of the control actions depending on the parameters of the object's movement and the parameters of the constraint plane. The proposed algorithms, in contrast to the traditional ones, are guaranteed to lead away from the obstacle regardless of its incline to the horizon.

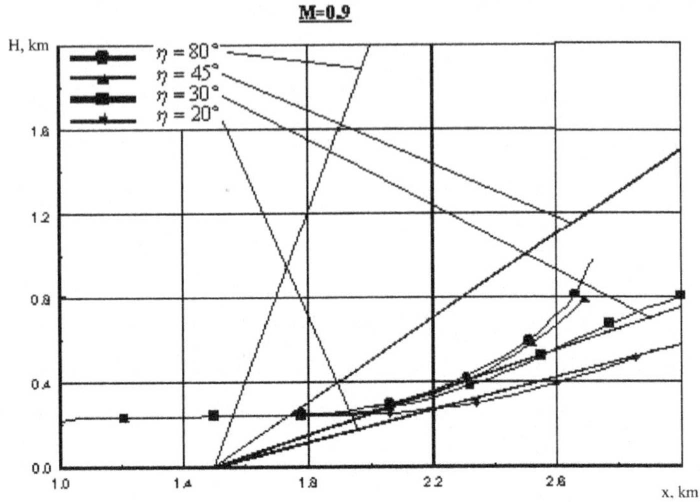

Fig. 5.17 Trajectories of the object movement during the maneuver of moving away from the inclined planes of restriction

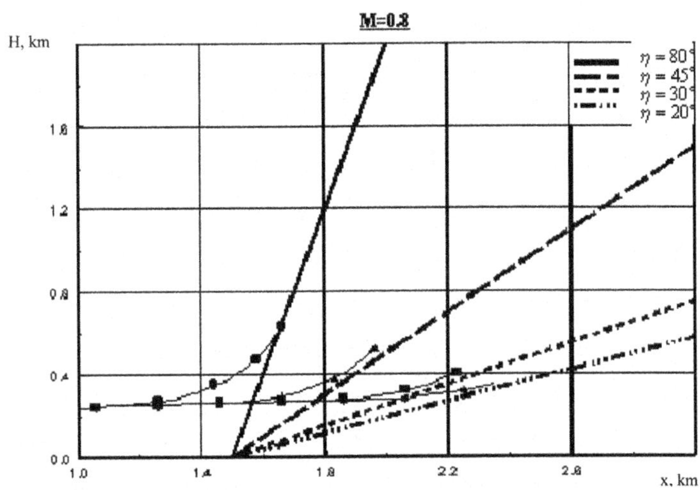

Fig. 5.18 Trajectories of the object movement during the maneuver of moving away from the inclined planes of restriction

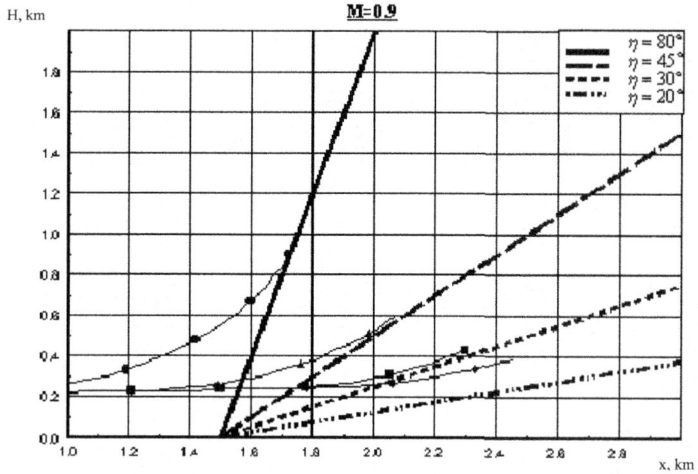

Fig. 5.19 Trajectories of the object movement during the maneuver of moving away from the inclined planes of restriction

References

1. Bukov VN (1982) Optimal algorithms in tasks with constraints of controlled coordinates. Publishing House of the USSR Academy of Sciences. Technical Cybernetics, No 2
2. Bukov VN, Ryabchenko VN, Kulabukhov VS, Naumov AI, Kosyanchuk VV, Goryunov SV (2001) Fundamentals of integration of aviation equipment systems.- Moscow: VATU
3. Byushgens GS, Studnev RV (1979) Aeroplane aerodynamics. Dynamics of longitudinal and lateral motion. - M.: Mashinostroenie
4. Mednikov VI (1976) Flight dynamics and aircraft piloting. Textbook.- Monino
5. Vygodsky M Ya (1958) Handbook of higher mathematics. - M.: State publishing house of technical and theoretical literature
6. Bukov VN (1987) Adaptive predictive flight control systems. - M.: Nauka
7. Akimov AN, Andreev VV (1997) Method of direct optimization. The main provisions and the order of application. - M: Air Force Engineering Academy, named by N.E. Zhukovsky

Chapter 6
Method and Algorithms for Direct Optimization of the Movement of a Damaged Aircraft

6.1 Analysis of the Main Prerequisites for Creating a Method of Direct Optimization

When solving problems of organizing the control of a damaged aircraft, the question very often arises: how sufficient are the remaining control resources to reach certain areas of space from the current or given state? If such areas are achievable, then what kind of trajectory does it have?

The situation can be significantly complicated by failures of controls or motors.

Therefore, the task of developing sufficiently simple algorithms, that allow solving trajectory analysis problems in real time, is very urgent.

In principle, the solution to such problems is possible, using methods of optimal control (OC).

A detailed analysis of each of the widely known approaches is an independent and rather complex scientific problem and is not included in the task of this book. This section discusses only those features, advantages, and disadvantages of optimization methods, that have found their continuation or served as the source of the proposed method of direct optimization of the movement of the AC.

For methods that implement the provisions of the theory of the classical calculation of variations (CCV) [1–6], the solution to the problem of trajectory analysis in the conditions of nonlinear differential connections and constraints seems to be practically unsolvable, especially in real time.

The principle of maximum of L. S. Pontryagin, as is well known, allows, in particular, to solve OC problems with nonlinear constraints in the form of inequalities. The abundance of areas of use and the results, obtained in them, indicate the high efficiency of this method [1, 2, 5, 7–12]. However, the same sources indicate that the numerical implementation of the principle of maximum in OC tasks faces problems caused by the limited resources of modern computers.

The need to simultaneously solve the main and conjugate systems of differential equations, which are mostly nonlinear, leads to a significant increase in the cost of machine time and the required reserves of RAM memory.

V. Akimov and V. A. Zatuchny, *Limiting Modes of Aircraft Flight*,
Springer Aerospace Technology, https://doi.org/10.1007/978-981-19-6329-2_6

So, for example, when the dimension of the matrix, determined by the main and conjugate systems, is doubled, the cost of machine time increases approximately 8 times, that is, in proportion to the cube of the increase in dimension [13].

The problem of the initial choice of the vector of conjugate variables in the theory principle of the maximum, which is analogous to the theory of CCV, also has no unambiguous solution. As before, the rate of convergence of the process of forming an optimal solution is largely determined by, how well the parameters of the vector of conjugate variables were set. Together with the need to solve the boundary value problem, the implementation of the principle of maximum requires a significant number of iteration cycles, and any minor change in the source data, as a rule, requires «reconfiguring» the parameters of numerical methods and resuming the solution almost «from zero». This disadvantage significantly limits the possibility of using the method in real-time problems.

The use of the dynamic programming method [14–18] for solving such problems in real time is simply impossible. This is due to the large dimension of the dynamical system and the complexity of the implementation of the mathematical apparatus of partial derivatives laid down in its basis.

The working out of direct variational methods [19] in the field of studying the dynamics of the flight of the AC makes their choice for solving the problems of trajectory analysis set out in this book the most preferable. However, the existing schemes for implementing these methods have a number of disadvantages that make it difficult to directly apply them to solve the problems formulated in the book. These include, first of all, the complexity of maintaining constraints on the functions of control and derivatives of phase coordinates.

The use of ADOR methods, despite the relatively low computational costs, in problems of trajectory analysis is difficult. This is due to the internal features and limitations of both the method itself and the mathematical model of the AC as an object of trajectory analysis.

Summarizing the above, we can conclude that in order to solve the problems of trajectory analysis set out in this book, as well as problems related to the synthesis of optimal controls of the spatial motion of the AC under the conditions of restrictions caused among them by failures and damages, the most preferable method would be the use of a method close in its idea to direct variational methods, but without their inherent disadvantages.

The method should:

a) provide a solution to the problem of trajectory analysis, that is, obtaining in one cycle of integration of a dynamic system a realizable basic trajectory of spatial motion AC that satisfies the current restrictions on the functions of control, the coordinates of the state vector, and their derivatives;

б) allow the solution of problems of optimal control in a certain sense of dynamic systems with a different number of circuits and channels of control, different from the number of phase coordinates;

в) provide a solution to the problem of optimization of the permissible basic trajectory under the conditions of restrictions on the functions of control, the coordinates of the state vector, and their derivatives;

г) to provide, in some specified cases, the identity, in the sense of the criteria, presented in p. 6.4, of the obtained optimal realizable permissible solutions to the solutions, found using the methods CCV and principle of the maximum of L.S. Pontryagin;

д) to ensure the time to obtain an acceptable realizable optimal solution is no longer than by existing methods for solving similar problems.

In this book, the authors offer their own viewpoints on solving the problems of trajectory analysis and optimal control, in particular, trajectory and flight dynamics problems. It is assumed that the optimal solution from the point of view of a given functional can be obtained as a result of the variation of the **derivative** of the permissible basic function of the coordinates of the state vector (CCV uses for this purpose directly the variation of the function of the coordinates of the state vector and the principle of maximum of L. S. Pontryagin—the variation of control).

As shown in p. 6. 2., the introduced new principle of forming optimal controls corresponds to the basic principle of the CCV theory, and at the same time, provides new opportunities in the process of forming the trajectory of the optimal permissible motion of the AC.

This principle eliminates the need for multiple integrations of direct and conjugate systems of nonlinear differential equations in the solution process. At the same time, obtaining the realized trajectory and controlling is carried out in one cycle of integration of the dynamic system.

The result of the development of the proposed approach was the creation of a method for direct optimization of the motion of dynamic systems. The main provisions of the method, a demonstration of its capabilities and a comparison of the results of the application with the known results of the existing methods are presented in this chapter.

6.2 A Loose Proof of the Correspondence of the Principle of Variation of the Derivative Function of an Arbitrary Curve to the Necessary Optimality Condition

In p. 6.1, it was determined that the method of direct optimization, developed in this section, is based on the principle of variation of the derivative functions of the state vector of the basic solution.

Let us consider how the introduction of this principle affects the nature of the optimal trajectories obtained as a result of its application. It can be assumed that, in the case of the immutability of a given space of solutions, a change in the type of functions, by which the differentiation of the objective function is carried out should not lead to a change in the type of optimal trajectories.

To prove this statement, we will use the following reasoning. It is known that the basis of the classical calculation of variations is the Euler equation. The derivation of this equation by the non-strict finite-difference method [3] is based on the assumption, accepted in mathematical analysis, that the extreme value of the functional for an arbitrary curve of a line can be obtained for this line based on its approximation by rectilinear links with given abscissas of their ends. In this case, the partial derivatives of the functional with respect to the ordinates of the vertex of the links must be equal to zero [3, 4].

Let us change this assumption in accordance with the basic principle of the developed method of direct optimization, namely: assume that the extremum of a given functional is obtained on a similar curve, on which its partial derivatives with respect to **the derivatives** of the function of the state coordinate are equal to zero.

Next, we apply the introduced assumption in the process of conclusion the necessary condition for the extremality of the functional under consideration. Comparing the obtained condition with the Euler equation, known from the CCV theory, it is obvious to assume that, in the case of their coincidence, the assumptions, accepted in their proof, are also adequate.

We will also conclude the necessary condition for the extremality of an arbitrary purpose function by a non-strict finite-difference method, similar to what was done in [3].

Consider the functional set in the form

$$J[x(t)] = \int_{t_{in'}}^{t_{fin}} F(t, x, \dot{x}) \cdot dt, \tag{6.1}$$

$x(t_{in}) = a$, $x(t_{fin}) = b$, t—independent variable, not necessarily time.

We will search for the optimal value of the functional (6.2) on polylines composed of a finite number of n rectilinear links with given abscissas of vertex (Fig. 6.1a)

$$t_{\text{in}} + \Delta t,\ t_{\text{in}} + 2 \cdot \Delta t, \ldots,\ t_{\text{in}} + (n-1) \cdot \Delta t, \tag{6.2}$$

where $\Delta t = (t_{\text{fin}} - t_{\text{in}}) \cdot n^{-1}$.

In this case, the functional $\boldsymbol{J}$ turns into a function of n−1 variables—x_1, x_2, …, x_{n-1}, and the problem of determining its extremum is reduced to the problem of finding such a set of these variables on which, in accordance with the previously accepted assumption, equality is achieved:

$$\partial J/\partial \dot{x}_1 = 0,\ \partial J/\partial \dot{x}_2 = 0,\ \ldots,\ \partial J/\partial \dot{x}_{n-1} = 0. \tag{6.3}$$

Solving the system (6.3), we find the values of x_i ($i = 1, \ldots, n-1$) at which the functional $\boldsymbol{J}$ can obtain an extreme value.

For the purpose of obtaining the necessary solution, we will perform the following operations. First of all, it is necessary to represent the value of the given functional (6.1) as an approximate integral sum

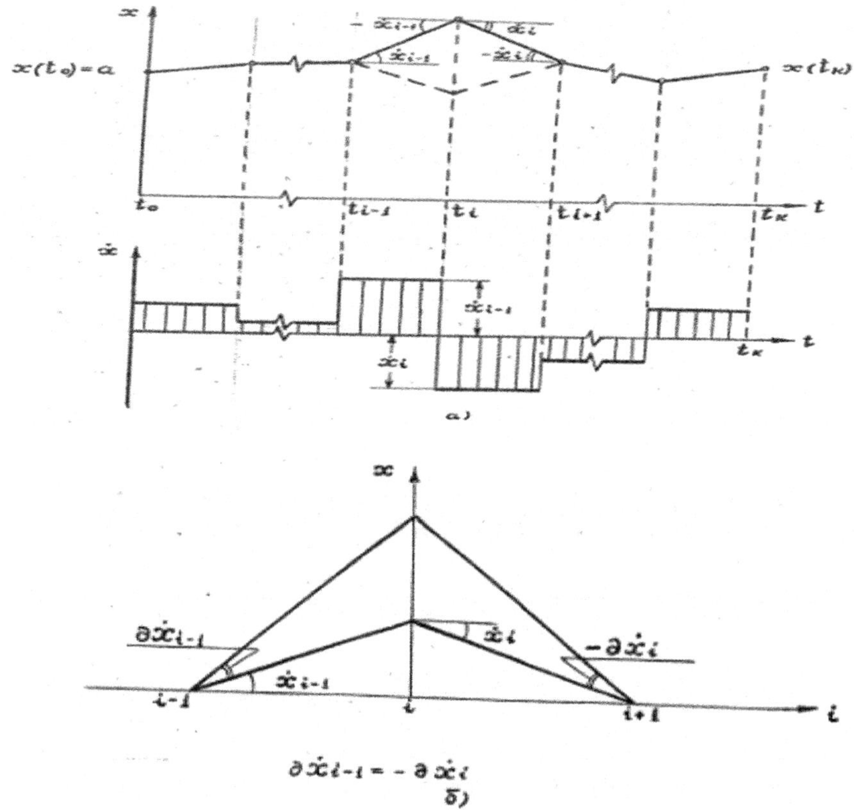

Fig. 6.1 Illustration of the proof of the Euler formula by the finite-difference method

$$J[x(t)] = \sum_{i=0}^{n} J_1[x_i(t)], \tag{6.4}$$

where

$$J_i[x_i(t)] = \int_{t_i}^{t_{i+1}} F\left(t_i, x_i, \frac{x_{i+1} - x_i}{\Delta t}\right) \cdot dt \approx F\left(t_i, x_i, {}^{\Delta x_i}/_{\Delta t}\right) \cdot \Delta t. \tag{6.5}$$

As a result of the replacement, the expression (6.4) can be converted to the form

$$J[x(t)] \approx \sum_{i=0}^{n} F\left(t_i, x_i, {}^{\Delta x_i}/_{\Delta t}\right) \cdot \Delta t. \tag{6.6}$$

Consider the partial derivatives of the functional $\boldsymbol{J}[x(t)]$ by independent variables $\dot{x}_i$. At the same time, special attention should be paid to the fact that in accordance with Fig. 6.1a, from the i-th derivative in the expression (6.6) depends only on two terms

$$J_i = F\left(t_i,\ x_i,\ \frac{x_{i+1} - x_i}{\Delta t}\right) \cdot \Delta t; \quad J_{i-1} = F\left(t_{i-1},\ x_{i-1},\ \frac{x_i - x_{i-1}}{\Delta t}\right) \cdot \Delta t. \quad (6.7)$$

Taking into account the made remark, we obtain an expression for the partial derivative of the original functional with respect to the variable $\dot{x}_i$ $(i = 1, 2, \ldots, n-1)$

$$\begin{aligned}
{}^{\partial J}\!/_{\partial \dot{x}_i} &= {}^{\partial J}\!/_{\partial \dot{x}_i} + \left({}^{\partial J}\!/_{\partial x_i}\right) \cdot \left({}^{\partial x_i}\!/_{\partial \dot{x}_i}\right) + \left({}^{\partial J}\!/_{\partial \dot{x}_{i-1}}\right) \cdot \left({}^{\partial \dot{x}_{i-1}}\!/_{\partial \dot{x}_i}\right) = \\
&= \left[\partial F\left(t_i,\ x_i,\ \frac{x_{i+1} - x_i}{\Delta t}\right)/\partial \dot{x}_i\right] \cdot \Delta t \\
&+ \left[\partial F\left(t_i,\ x_i,\ \frac{x_{i+1} - x_i}{\Delta t}\right)/\partial x_i\right] \cdot (\partial x_i/\partial \dot{x}_i) \cdot \Delta t + \\
&+ \left[\partial F\left(t_{i-1},\ x_{i-1},\ \frac{x_i - x_{i-1}}{\Delta t}\right)/\partial \dot{x}_{i-1}\right] \cdot (\partial \dot{x}_{i-1}/\partial \dot{x}_i) \cdot \Delta t = \\
&= F_{\dot{x}_i}\left(t_i,\ x_i,\ \frac{x_{i+1} - x_i}{\Delta t}\right) \cdot \Delta t + F_{x_i}\left(t_i,\ x_i,\ \frac{x_{i+1} - x_i}{\Delta t}\right) \cdot (-1) \cdot \Delta t + \\
&+ F_{\dot{x}_{i-1}}\left(t_{i-1},\ x_{i-1},\ \frac{x_i - x_{i-1}}{\Delta t}\right) \cdot (-1) \cdot \Delta t. \qquad (6.8)
\end{aligned}$$

Reducing the terms of the final expression (6.8) by–Δt^2 and equating the found derivative to zero, we get the following expression:

$$\begin{aligned}
&F_{x_i}(t_i,\ x_i,\ \Delta x_i/\Delta t) - \frac{F_{\dot{x}_i}(t_i,\ x_i,\ \Delta x_i/\Delta t) - F_{\dot{x}_{i-1}}(t_{i-1},\ x_{i-1},\ \Delta x_{i-1}/\Delta t)}{\Delta t} = 0, \\
&or\ F_{x_i}(t_i,\ x_i,\ \Delta x_i/\Delta t) - \Delta F_{\dot{x}_i}/\Delta t = 0, \quad (i\ =\ 1, 2, \ldots, n-1)
\end{aligned} \qquad (6.9)$$

The resulting expression needs a number of explanations, regarding some of the substitutions, made during the transformations. In particular, when the final expression (6.8) was concluded, the following substitutions were made:

$$\partial x_i/\partial \dot{x}_i = -\Delta t, \qquad (6.10)$$

and also

$$\partial \dot{x}_{i-1}/\partial \dot{x}_i = -1. \qquad (6.11)$$

The correctness of the formula (6.10) is obvious and can be easily illustrated by the example of Fig. 6.1a, in which the increment of the coordinate of the basic trajectory and its derivative differ in sign, and their ratio is proportional to the integration step.

Below, in Fig. 6.1b, it is shown that the substitution (6.11) also has a geometric meaning. Due to the fact that only the position of the i-th point of the trajectory varies at each time t_i, and the points corresponding to time t_{i-1} and t_{i+1} are fixed, an arbitrary increment of the derivative $\partial\,\dot{x}_{i-1}$ necessarily causes an equal in modulus and opposite in sign increment of the derivative $\partial\,\dot{x}_i$.

Let us return to the last result obtained in (6.9). The solution of this system of equations with respect to $x_i\,(i = 1, 2, \ldots, n-1)$ allows you to define a set of values $X = \{x_1, x_2, \ldots, x_{n_1}\}$, on which the functionality (6.1) can take an extreme value.

Directing the number of partitions of n to infinity, we get an infinite sequence $X^* = \{x_1, x_2, \ldots\}$, which can also be matched by an infinite sequence $\{J_j\}$. If the limit of this sequence exists, then it can be the desired extremum of the functional (6.1).

Assume that the functions F and $\dot{x}$, considered above had the necessary properties of continuity and differentiability with respect to the corresponding arguments. Then, in the last Eq. (6.9), a limit transition is possible at $n \to \infty$, the result of which is an equation that completely coincides with the Euler equation

$$F_x(t, x, \dot{x}) - d[F_{\dot{x}}(t, x, \dot{x})]/dt = 0. \tag{6.12}$$

Thus, equality (6.12) fully confirms the correctness of the assumption made at the beginning of this section about the adequacy of the approaches of the CCV theory and the developed method of direct optimization. Therefore, the principle of variation of the derivative of the basic solution can be used to construct the theory of a new method.

The main provisions of this method are presented below.

6.3 Development of the Main Provisions of the Method of Direct Optimization of the Motion of Dynamic Systems

6.3.1 *General Formulation of Optimal Control Problems and Determination of the Sequence of their Solution in the Method of Direct Optimization of Dynamic Systems*

We will consider an arbitrary object of study, the behavior of which is described by some system of n ordinary nonlinear differential equations

$$\dot{x} = f(x, u, t), \tag{6.13}$$

where $x \in R^n$—tate vector; $U \in R^m$—vector of control, included in the general case in the right part of the system (6.13) nonlinear; t—time, belonging to the given interval $[t_{\text{in}}, t_{\text{fin}}]$; $f(\cdot)$ —vector n-dimensional nonlinear function.

In general, it is assumed, that for time values t_{in} and t_{fin} are given corresponding initial and final conditions

$$x(t_{\text{in}}) = x_{\text{in}},\ x(t_{\text{fin}}) = x_{\text{fin}},\ u(t_{\text{in}}) = u_{\text{in}},\ u(t_{\text{fin}}) = u_{\text{fin}} \tag{6.14}$$

In this case, only those values of the phase coordinates and controls that belong to the corresponding closed areas of their permissible values G_x and G_u, represented as constraints of the inequality type, are considered.

In the considered conditions, the task is set to determine such a vector of controls *u(t)* and the corresponding vector of phase coordinates *x(t)*, which would deliver an extreme value to some fixed functional *W(x, u, t)*, the specific type and physical meaning of which is determined by the nature of the object under study

$$W(x, u) \xrightarrow[\{G_x,\ G_u\}]{} \inf\ (\sup). \tag{6.15}$$

Other settings of the task are also possible, for example, in terms of giving initial and final conditions. So, in tasks with moving ends is required to observe at the ends of the integration segment the constraints of such type

$$\Psi_{\text{in}}(x) = 0,\ \Psi_{\text{fin}}(x) = 0, \tag{6.16}$$

where $\Psi_{\text{in}}(x) = 0$ or $\Psi_{\text{fin}}(x) = 0$ represent the equations of surfaces.

In tasks for operating speed, the time point t_{fin} is considered as the purpose function.

Let us consider the main ideas of applying the developed method of direct optimization to solving problems of optimal control.

To do this, we analyze the scheme for solving the problem of optimizing the movement of the object (6.13) in the sense of finding trajectories and controls, that deliver an extreme value to the functional (6.15).

It is known [7] that the extremum of the functional corresponds to the equality of the linear part of the variation of the functional to zero

$$\delta W = \frac{\partial W}{\partial x^T}\delta x + \frac{\partial W}{\partial u^T}\delta u = 0, \tag{6.17}$$

where δx and δu are the corresponding variations of state and control.

Since x and u are connected by Eq. (6.13), the auxiliary functional is used to solve the problem

$$W = W(x, u) + \int_{t_{in'}}^{t_{\text{fin}}} \lambda^T(t) \cdot [\dot{x} - f(x, u, t)]dt, \tag{6.18}$$

where the second term takes into account the differential relations; $\lambda^T(t) \in R^n$ is a vector, whose components are yet undefined functions called Lagrange multipliers.

Using these multipliers, the task about the conditional extremum of the functional (6.15) is reduced to the task of the unconditional extremum of the functional (6.13), which corresponds to the (2n + m) Euler–Lagrange system. As a result of their solutions, extreme values of *x(t)* and *u(t)* were obtained.

However, a different view of the solution to the functional minimization problem is possible (6.15). To do this, consider the equation of the relation of variations δx and δu

$$\delta \dot{x} = \frac{\partial t}{\partial x^T}\delta x + \frac{\partial t}{\partial u^T}\partial u. \tag{6.19}$$

Based on the previous proof, we take the vector $\delta \dot{x}$as independent variations, then expressing δu from (6.19), we get [20]

$$\begin{aligned} \delta u =& \frac{\partial f^{\#}}{\partial u^T} \cdot \left(\delta \dot{x} - \frac{\partial f}{\partial x^T}\delta x\right); \\ \delta x =& \delta \dot{x} \cdot \Delta t, \end{aligned} \tag{6.20}$$

where the second equation is valid by virtue of Eqs. (6.13) or (6.19) when they are integrated on the interval Δt with the control (6.20).

So, instead of two independent variations δx and δu, we have one $\delta\ \dot{x}$.

Substitute the expression for δu (6.20) in the second term of the functional variation expression (6.17). After making the appropriate changes, taking into account the equality of

$$\frac{\partial W}{\partial u^T} \cdot \frac{\partial t^{\#}}{\partial u^T} = \frac{\partial W}{\partial f^T} = \frac{\partial W}{\partial \dot{x}^T} \, and \, \frac{\partial W}{\partial u^T} \cdot \frac{\partial t^{\#}}{\partial u^T} \cdot \frac{\partial t}{\partial x^T} = \frac{\partial W}{\partial f^T} \cdot \frac{\partial t}{\partial x^T} = \frac{\partial W}{\partial x^T},$$

we obtain the necessary extremum condition in the form

$$\delta W^* = \frac{\partial W}{\partial \dot{x}^T}\delta \dot{x} = 0, \tag{6.21}$$

where $\delta\ \dot{x}$ corresponds to (6.19).

Thus, the same situation is observed as in the case of conditional minimization of the function of two variables, connected by the equation of connection, when the method of exclusion of variables is used instead of the Lagrange multiplier method [21]. Its essence is to express one variable through another, using the equation of connection. As a result, the task is reduced to optimizing the function, but only one variable.

A similar situation occurs in the case under consideration.

Note specifically that the functional which depends directly only on *x* and *u*, due to the presence of the equation of connection (6.13), turns out to be dependent on

$\dot{x}$. Thus, by performing a variation of the derivatives, one can immediately solve the problem for the conditional extremum, without reducing it, as in CCV, to the problem for the unconditional extremum for the functional (6.18).

It should be emphasized once again that the variations δx and δu, unlike CCV, are not arbitrary, but are related to Eq. (6.19) with variation $\delta \dot{x}$.

Consider the set of Eqs. (6.13) and (6.20). It can be noted that the sequential solution of Eqs. (6.20) and (6.13) allows us to form a basic solution. In this case, the type of this solution and the nature of the resulting trajectory of the dynamic object will be completely determined by the form of the derivative assignment $\dot{x}$ in Eq. (6.20). Obviously, the function $\dot{x}(t)$ can be set, so that it first provides a solution to the original boundary problem and, for second, it delivers an extreme value to the functional (6.15). In fact, this leads to the fact that the function of changing the derivative $\dot{x}$(t) in this problem acts as the main variable parameter.

The presented form of organization of the process of forming solutions to OS problems has the following advantages.

First, applying the pseudo-circulation operation to calculate the matrix $[\partial f/\partial u^{\mathrm{T}}]^{\#}$ in the expression (6.20) allows for solving OS problems for dynamic systems with an arbitrary number of channels of control. In this case, you can get, depending on the dimension of the matrix $[\partial f/\partial u^{\mathrm{T}}]$ and its rank, three cases:

1. In case m $= n$ and $rk[\partial f/\partial u^{\mathrm{T}}] = n$ vector of controls, calculated by (6.20), vector of controls, calculated by, when substituting in (6.13) provides exact equality of the right and left sides of the given equation. In this case, the «defect» of the variation, defined by the expression $\Delta W = \delta W - \delta W*$, is equal to zero.
2. In case m $< n$ or $rk[\partial f/\partial u^{T}] < n$ vector of controls, calculated by (6.20), using the pseudo-circulation operation, provides the minimum value of the norm [22]

$$\|\dot{x} - f(x,u,t)\| \rightarrow min, \quad \Delta W = \frac{\partial W}{\partial u}\left(E - \frac{\partial f^{\#}}{\partial u}\cdot\frac{\partial f}{\partial u}\right)\delta u.$$

3. In case m $> n$ and $rk[\partial f/\partial u^{\mathrm{T}}] = n$ vector of controls, calculated in accordance with (6.20), also minimizes the same norm and simultaneously takes a value, that minimizes its own norm $u \rightarrow min$ [22]. Here and further, the Euclidean norm is taken into consideration [10]. In this case, the «defect» of the functional's variation is $\Delta W = 0$.

Second, and this will be shown in the following paragraphs, setting the vector of derivatives $\dot{x}(t)$ in the form of a mismatch function between the current and final values of the phase coordinates allows for one cycle of integration of the dynamic system to obtain a basic solution to be implemented. At the same time, the procedure for maintaining the given constraints for control, state coordinates, and their derivatives is significantly simplified.

In addition, it should be noted that, as will be shown below, in comparison with the maximum principle of L.S. Pontryagin, the dimension of the solved dynamical system is reduced by half (since there are no equations of the conjugate system). In the case of solving CCV problems, there is no need to search for Lagrange multipliers and, accordingly, determine the extremum of the functional (6.19).

Let's consider, how the demonstrated principle is implemented for solving more complex OS problems, which are the solutions to the proposed method of direct optimization.

In general case, the solution of the set task of the optimal control is proposed to be found in the form of a sequence of the following operations:

- the mathematical formulation of the original task of the optimal control is reduced to a form that is convenient for implementing the method of direct optimization;
- a level-by-level decomposition of the mathematical model of the control object is performed;
- analytical dependencies of expressions for calculating control functions at the corresponding sublevels of the original model are formed;
- procedures for maintaining constraints on state and control vectors are defined;
- an algorithm for constructing an acceptable and realizable basic solution to aboundary problem is formed;
- a variation of the basic solution in accordance with the requirements of ensuring the extreme value of the purpose function is performed.

The set of the presented operations forms a system of the main provisions of the developed method of direct optimization, their consideration is carried out in the following subsections.

6.3.2 Level-by-Level Decomposition of the Mathematical Model of the Control Object

This procedure is necessary for objects, whose controls are included only in the part of the equations, describing the movement of the object of study, and the requirements for the final state or optimization criterion depend on the entire set of components of the state vector.

The level-by-level decomposition of the mathematical model of the control object includes the following operations.

Let the initial mathematical model (MM) have the form (6.13). The first (lower) level of MM is obtained based on the analysis of the matrix n × m of the form

$$\left[\frac{\partial f}{\partial u^T}\right] = \begin{bmatrix} * & * & \cdots & * \\ 0 & 0 & \cdots & 0 \\ - & - & - & - \\ * & * & * & * \\ 0 & 0 & 0 & 0 \end{bmatrix} \begin{matrix} (1) \\ (1) \\ \cdots \\ (n-1) \\ n \end{matrix} \qquad (6.22)$$

$$(1) \quad (2) \ldots \quad (m),$$

which is obtained by differentiating the vector function $f(x, u, t)$ by the control vector u.

Let's assume that in (6.22) and beyond, k elements of the matrix, other than zero, are marked with asterisks. Then the lower level of MM will be represented by a k-dimensional system of equations

$$x^{\mathrm{k}} = f^k(x^k, u, t), \tag{6.23}$$

where $x^k = [x^1, x^2, \ldots, x^k]^T$, $f^k = [f^1, f^2, \ldots, f^k]^T$.

The remaining without these equations model of the first level, consisting of (n-k) equations, is written as

$$x^{n-k} = f^{n-k}(x^{n-k}, x^k, t) \tag{6.24}$$

This model is differentiated in a similar way by variables x^k, which are the controls for this level. As a result, a second-level mathematical model is formed, in which the number of equations and the coordinates, defined by them, is equal to the number of non-zero derivatives of the system (6.24) with respect to the variables x^k. Let this number be $l \leq n-k$, then MM of the second level is written as

$$x^l = f^l(x^l, x^k, t). \tag{6.25}$$

The remaining model, where the x^l vector acts as the control, will have the following form:

$$x^{n-k-l} = f^{n-k-l}\left(x^{n-k-l}, x^l, t\right). \tag{6.26}$$

Suppose that in the case of differentiating a vector function f^{n-k-l} by x^l a matrix is obtained, that has non-zero elements in all strings (at least one)

$$\left[\frac{\partial f^{n-k-l}}{\partial x^l}\right] = \begin{bmatrix} * & * & \cdots & * \\ \cdots & \cdots & \cdots & \cdots \\ * & * & \cdots & * \end{bmatrix} \begin{matrix} (1) \\ \cdots \\ n-k-l \end{matrix} . \tag{6.27}$$
$$\quad (1) \quad (2) \ldots (l)$$

This is the end of the MM decomposition procedure. Otherwise, it should be continued, until a matrix without zero strings is obtained, similar to (6.27).

As a result of the decomposition operation, carried out in this way, a scheme of multi-level controllability by the coordinates of the state vector of the original dynamical system is obtained (6.13), namely: the control u directly affects the change in the derivatives of the k coordinates of the state vector -x^k. They, in turn, are «controls» and affect the change in the derived ℓ coordinates of the original state vector x^ℓ, which then themselves act as the control actions of the remaining ($n-k-\ell$) part of the coordinates.

If we consider the proposed scheme, starting from the upper level of the decomposition, we can see, and this will be shown below, that the formation of the derived coordinates of the state vector at the considered level leads to the determination of the required control actions of a given dynamic system at lower levels. It is obvious that, in the end, this approach will allow you to form directly the vector of controls u.

6.3.3 *The Main Provisions of the Method of Direct Optimization*

6.3.3.1 Equations of Movement of the Research Object

In accordance with the general statement of the problem in this book, we will consider the developed method of direct optimization by the example of solving specific problems of controlling the movement of the AC under the conditions of restrictions. At the same time, we will distinguish two levels of complexity of the tasks to be solved.

First, these are the problems of trajectory analysis, which in this paper is understood as an assessment of the possibility of reaching certain areas of space by a damaged AC. The object of study (in this case, the AC) is considered as a material point of variable mass. It is also assumed that there are no turbulent perturbations of the atmosphere and its parameters are stationary.

Second, these are the tasks of control synthesis in the pilot circuit, when the forces and moments, acting on it, are introduced into the description of the movement of the initial object of research, which are directly related to the influence of the various controls available on board of the aircraft. The need to solve the problem is due to an attempt to include a rotary thrust vector (RTV) in the composition of the controls of the damaged AC.

Let us consider the solution to the problem of optimal control of the spatial motion of the AC as an object of trajectory analysis. We will use the differential equations of the spatial motion of the AC under the action of these forces obtained for these conditions in [23]. In projections on the axis of the trajectory coordinate system, these equations have the form

$$\dot{V} = \left[P \cdot \cos\left(\alpha - \phi^{*}\right) - x_{f} - m \cdot g \cdot \sin\theta\right]/m; \tag{6.28}$$

$$\dot{\theta} = \left[P \cdot \cos\gamma \cdot \left(\sin\left(\alpha - \phi^{*}\right) + Y_{f}\right) - m \cdot g \cdot \cos\theta + m \cdot V^{2} \cdot \cos\theta\right]/(m \cdot V); \tag{6.29}$$

$$\dot{\varphi} = -\left[P \cdot \sin\gamma \cdot \left(\sin(\alpha - \varphi^{*}) + Y_{f}\right)\right]/(m \cdot V \cdot \cos\theta). \tag{6.30}$$

Here and after, all designations are accepted in accordance with [24].

To the equations of motion, you need to add kinematic equations that connect the lateral, longitudinal range *Z, L,* and height *H* with the speed of flight V and the angles θ and φ

$$\dot{L} = V \cdot \cos\theta \cdot \cos\varphi; \tag{6.31}$$

$$\dot{H} = V \cdot \sin\theta; \tag{6.32}$$

$$\dot{Z} = -V \cdot \cos\theta \cdot \sin\varphi, \tag{6.33}$$

as well as the equation that characterizes the change of the mass of the AC over time due to fuel burn up

$$\dot{m} = -q_c, \tag{6.34}$$

where q_c—second fuel consumption AC, kg/sec.

Obviously, the Eqs. (6.28)–(6.33) can in many cases be considered separately from (6.34), forming a system of six independent equations, containing 9 unknowns: $V,\ P,\ \alpha,\ \gamma,\ \varphi,\ \theta,\ L,\ H,\ Z.$

Since the system is not closed and there are three «extra» unknowns, it is necessary to select those that could be used as functions of control. In principle, all variables are equal, and any of them can be selected as part of the vector of control. You can notice that the variables *L, H, Z, V,* θ and φ are included in the Eqs. (6.28)–(6.33) together with their first derivatives and characterize the state of the object at each time *t*; their number is equal to the order of the system. Simultaneously. the values of *P,* α, γ are contained in the same equations without their derivatives and play the role of free variables. They can be defined as some functions of time and determine the decision of the system (the behavior of the control object) at a given initial state *L, H, Z, V,* θ, *and* φ. Therefore, it is advisable to choose the functions α(*t*), $\gamma(t)$, and *P*(*t*) as the control functions.

In this case, the vector of control will take the form

$$u(t) = [\alpha(t), \gamma(t),\ P(t)]^{\mathrm{T}}. \tag{6.35}$$

Variables, that are not explicitly included in the controls, form a vector of phase coordinates

$$F(t) = [x(t),\ y(t)]^{\mathrm{T}}, \tag{6.36}$$

where

$$\begin{cases} x(t) = [L(t),\ H(t),\ Z(t)]^T, \\ y(t) = [V(t),\ \theta(t),\ \phi(t)]^T. \end{cases} \tag{6.37}$$

Its components, as well as the components of the vector *u(t)*, must satisfy the imposed boundary conditions of the original task. This issue is discussed in more detail in Sect. 6.3.4.

We will use the decomposition operation the considered MM of the controlled motion of the AC, introduced earlier in the Sect. 6.3.2. As a result of its execution, the original MM can be represented as the following dynamic system:

$$\begin{cases} \dot{x} = f(x,\ y,\ t), \\ \dot{y} = \Psi(y,\ u,\ t), \end{cases} \quad t \in [t_{\text{in}},\ t_{\text{fin}}] \tag{6.38}$$

where vectors f and Ψ, formed by sets $\{f_1, f_2, f_3\}$ and $\{\Psi_1, \Psi_2, \Psi_3\}$ accordingly.

The initial conditions of the process under study are set by the values of the phase coordinates and controls

$$x(t_{\text{in}}) = x_{\text{in}} = [L_{\text{in}},\ H_{\text{in}},\ Z_{\text{in}}]; \tag{6.39}$$

$$y(t_{\text{in}}) = y_{\text{in}} = [V_{\text{in}}, \theta_{\text{in}}, \varphi_{\text{in}}]; \tag{6.40}$$

$$u(t_{\text{in}}) = u_{\text{in}} = [P_{\text{in}}, \alpha_{\text{in}}, \gamma_{\text{in}}]. \tag{6.41}$$

The corresponding values at the time t_{fin} make up the block of end condition

$$x(t_{\text{fin}}) = x_{\text{fin}} = [L_{\text{fin}},\ H_{\text{fin}},\ Z_{\text{fin}}]; \tag{6.42}$$

$$y(t_{\text{fin}}) = y_{\text{fin}} = [V_{\text{fin}}, \theta_{\text{fin}}, \varphi_{\text{fin}}]; \tag{6.43}$$

$$u(t_{\text{fin}}) = u_{\text{fin}} = [P_{\text{fin}}, \alpha_{\text{fin}}, \gamma_{\text{fin}}]. \tag{6.44}$$

The set of the dynamic system (6.38) and the given boundary conditions (6.39)–(6.44) leads to the boundary value task. At the same time, a mandatory additional requirement for its solution is the need to comply with the current system of restrictions, determined by the failures, that occur during the flight.

The set of solutions of the original problem that satisfy all the given conditions forms the set of acceptable solutions of D.

We will assume that each permissible control *u(t)* of *D* is evaluated by the value of some purpose function *W*, defined at the right ends of the permissible trajectories. Control is assumed as optimal if it delivers the extreme value of the W function. Its specific type is determined by the given indicator of efficiency.

6.3.3.2 Formation of a Difference Scheme for Approximating the Basic Trajectory of the Damaged Aircraft

Consider a dynamic system (6.38). It is known from the theory of mathematical analysis [9, 10, 25, 26] that this system forms in a convex closed domain of solutions, bounded, for example, by the intersection of the cube of boundary conditions (6.39)–(6.44) with a set of $\{G_x, G_u\}$, the set of continuous functions of changing the state coordinates, belonging to the class of curves KC^2.

Obviously, this set is finite. Therefore, the problem of forming an acceptable basic solution to the initial dynamical system can be reduced to the task of finding such function of changing the coordinates of the state $x(t) \in KC^2$, which at $t \in [t_{in}, t_{\text{fin}}]$ took values from the set G_x, and at the ends of the segment $[t_{\text{in}}, t_{\text{fin}}]$— values of known boundary conditions (6.39)–(6.44). Any solution, that satisfied these requirements can be selected as a basic solution.

The nature of the trajectories, implemented by the damaged AC, as a rule, is reduced to the organization of movement along the shortest distance to specific areas of space with minimal required resources, that is, «in a straight line». This is due to quite understandable reasons (limited controllability, the risk of further aggravation of the consequences of damage or failure, etc.). In this regard, this book deliberately narrows the space of possible trajectories on which the extremum of the AC functional is sought.

Since one of the main requirements for the desired solution is the fulfillment of the specified boundary conditions at the ends of the segment $[t_{\text{in}}, t_{\text{fin}}]$, then at least starting from some point in time $t \in [t_{\text{in}}, t_{\text{fin}}]$, we will form a function for targeting the state coordinate vector $x(t)$ based on the principle of minimizing the square of the discrepancy between the final and current values of the given coordinates. To this end, we write the expression for the square of the residual in the form

$$\rho(u) = (x_k - x(t + \Delta t))^{\text{T}}(\text{x}_k - \text{x}(\text{t} + \Delta\text{t})), \tag{6.45}$$

where the value of the current coordinate is approximated as Taylor's row with an order accuracy $O(\Delta t^2)$

$$x(t + \Delta t) = x(t) + \dot{x}(t) \cdot \Delta t + \ddot{x}(t + \Delta t) \cdot \Delta t^2/2 + O\left(\Delta t^2\right). \tag{6.46}$$

We will look for the control $u(t)$ as a function that minimizes the current value of the function $\rho(u)$ at each given point of time $t \in [t_{\text{in}}, t_{\text{fin}}]$. To do this, we use the necessary condition of its extremum in the form

$$\partial\rho(u)/\partial u = 0 \tag{6.47}$$

Due to the fact that

$$\ddot{x}(t + \Delta t) = [\partial f/\partial x]\cdot f(x, y, t) + [\partial f/\partial y] \cdot \Psi(y, u, t), \tag{6.48}$$

we will write

$$\frac{\partial\rho(u)}{\partial u} = 2\cdot\left(x_h - x(t) - \dot{x}(t)\cdot\Delta t - \ddot{x}(t+\Delta t)\cdot\Delta t^2/2\right)\cdot(-[\partial f/\partial y]\cdot[\partial\Psi/\partial u]) = 0, \tag{6.49}$$

where do we get that

$$\ddot{x}(t+\Delta t) = 2\cdot\left(\frac{x_h - x(t)}{\Delta t^2} - \frac{f(x,\ y,\ t)}{\Delta t}\right). \tag{6.50}$$

We substitute in (6.50), instead of the second derivative $\ddot{x}(t+\Delta t)$, its component from (6.48), which depends on the control

$$[\partial f/\partial y]\cdot\Psi(y,\ u,\ t+\Delta t) = 2\cdot\left(\frac{x_h - x(t)}{\Delta t^2} - \frac{f(x,\ y,\ t)}{\Delta t}\right); \tag{6.51}$$

then it will be possible to write that

$$\Psi(y,\ u,\ t+\Delta t) = [\partial f/\partial y]^{\#}\cdot 2\cdot\left(\frac{x_h - x(t)}{\Delta t^2} - \frac{f(x,\ y,\ t)}{\Delta t}\right). \tag{6.52}$$

Note that this expression is obtained for the model of the trajectory motion of the AC (6.28)–(6.33), for which the matrix $[\partial f/\partial x]$ is zero. For other objects, (6.52) will differ by the presence of the corresponding term.

In the case, where an explicit expression for the function of control vector, can be obtained from Eq. (6.52), the problem of forming a basic solution, that minimizes the residual norm (6.45), can be solved. However, the prevailing number of tasks does not allow the explicit representation of the expression for *u(t).* In this case, it is possible to apply an iterative approach to the synthesis of the control. However, you can use another method, for example, a difference scheme for determining the necessary control.

Decomposing the left part of the expression (6.52) into a row and holding the terms of the first order of smallness, we get the following expression:

$$\Psi(y,\ u,\ t+\Delta t) = \Psi(y,\ u,\ t) + [\partial\Psi/\partial u]\cdot\Delta u + [\partial\Psi/\partial y]\cdot\Delta y, \tag{6.53}$$

from which follows

$$\Delta u = \left[\frac{\partial\Psi}{\partial u}\right]^{\#}\cdot\left(\left[\frac{\partial f}{\partial y}\right]^{\#}\cdot 2\cdot\left(\frac{x_h - x(t)}{\Delta t^2} - \frac{f(x,\ y,\ t)}{\Delta t}\right) - \Psi(y,\ u,\ t) - [\partial\Psi/\partial y]\cdot\Delta y\right) \tag{6.54}$$

Here, the «#» symbol is the symbol for the pseudo-conversion operation of the matrices of partial derivative $[\partial f/\partial y]$ and $[\partial\Psi/\partial u]$, described in [22].

In order to simplify the writing of the formula (6.54), we introduce a new notation of the form

$$\Delta y_{\bullet} = 2 \cdot \left[\frac{\partial f}{\partial y}\right]^{\#} \cdot \left(\frac{x_h - x(t)}{\Delta t} - f(x,\ y,\ t)\right), \tag{6.55}$$

Where Δy_n is the required increment of the «pseudo-control» vector for the first equation of the system (6.38).

Taking into account the last formula, we obtain the following system of difference equations, which allows us to calculate the necessary control, that minimizes the current discrepancy of the state coordinates (6.45)

$$\begin{cases} \Delta y_{\bullet} = 2 \cdot \left[\frac{\partial f}{\partial y}\right]^{\#} \cdot \left(\frac{x_h - x(t)}{\Delta t} - f(x,\ y,\ t)\right), \\ \Delta \mathrm{u} = \left[\frac{\partial \Psi}{\partial u}\right]^{\#} \cdot \left(\frac{\Delta y_{\bullet}}{\Delta t} - \Psi(y,\ u,\ t) - [\partial \Psi / \partial y] \cdot \Delta y\right). \end{cases} \tag{6.56}$$

It is obvious that the controls, formed as a result of the solution of the system (6.56), may well provide such a trajectory of motion of a dynamic object, which is guaranteed will start and end at points with the coordinates of the state vector, determined by the given boundary conditions. At the current time t in the second expression (6.56) $\Delta y = 0$. Then, at the interval of integrating the trajectory during the time of cycle, the resulting control Δu is corrected by the value

$$\Delta u(\Delta y) = -\left[\frac{\partial \Psi}{\partial u}\right]^{\#} \frac{\partial \Psi}{\partial y} \Delta y. \tag{6.57}$$

In order to finally define the difference scheme obtained in (6.56), and make it suitable for calculations, it is necessary to make the following remarks about the space of solutions given by this scheme. For this purpose, we will move from considering the space of continuous functions to the space of grid functions, constructed on a grid with a fixed step Δt (in general case, the division can be carried out not by time, but by any of the coordinates of the vector $x(t)$). In this case, continuous functions $x(t)$, $y(t)$, and $u(t)$ will be approximated by their grid equivalents $x(t_i)$, $y(t_i)$, *and* $u(t_i)$, where still the same $t_i \in [t_{in},\ t_{\text{fin}}]$.

In the new space under consideration, the values of the grid functions $x(t_i)$, $y(t_i)$, *and* $u(t_i)$ will be determined sequentially, using the expressions (6.56) and (6.38), namely: the difference scheme (6.56) is the source of the formation of the current increments of control functions at a time t_{i+1}, and system (6.38) is the source of formation of the phase coordinates corresponding to the found control. In this case, the expression Δt should be understood as the notation of the difference t_{fin}-t, and then

$$T_{fin} - t_i = \Delta t_i. \tag{6.58}$$

Initial values of vector components $f(x_i, y_i, t_i)$ and $\Psi(y_i, u_i, t_i)$ are considered to be known from the boundary conditions (6.39)–(6.44) and can be expressed as shown in [19].

Thus, the working version of the system (6.56), which determines the increment of the control vector $\Delta u_{i\text{fin}}$ in the case of integrating the system (6.38) «from left to right» (that is, from the moment of time t_{in} to moment of time t_{fin}) in the space of grid functions will look like this

$$\begin{cases} \Delta y_{ih} = 2 \cdot \left[\frac{\partial f}{\partial y}\right]_i^{\#} \cdot \left(\frac{x_h - x(t_i)}{\Delta t_i} - f(x_i,\ y_i,\ t_i)\right); \\ \Delta u_{ih} = \left[\frac{\partial \Psi}{\partial u}\right]_i^{\#} \cdot \left(\frac{\Delta y_{ih}}{h(t_i)} - \Psi(y_i,\ u_i,\ t_i)\right), \end{cases} \quad (6.59)$$

where $h(t_i)$—in general form, an arbitrary time function, that determines the intensity of the generated controls. A detailed discussion of this will be given below.

As already mentioned, the implementation of the scheme (6.59), due to the method of its preparation, guarantees the convergence of the parameters of the basic trajectory to the values of the boundary conditions in the coordinates of the state vector. If the conditions of the problem do not require convergence in the values of the derivatives of the state vector and controls, then the procedure for forming a basic solution, defined by the system (6.59), can be considered complete.

In all other cases, the solution to the boundary value problem requires the introduction of an additional system of differences that sets the increments of the control functions for integrating the dynamical system (6.38) «from right to left», that is, from the moment of time t_{fin} до момента t_{in}. For this purpose, the values of the vectors $f(x_i, y_i, t_i)$ and $\Psi(y_i, u_i, t_i)$ in Eq. (6.59) are sufficient to replace with values $f(x_{\text{fin}}, y_{fin}, t_{\text{fin}})$ and $\Psi(y_{\text{fin}}, u_{fin}, t_{fin})$, and matrices of partial derivative $[\partial f/\partial y]_i^{\#}$и $[\partial \Psi/\partial u]_i^{\#}$ to replace with values $[\partial f/\partial y]_k{}^{\#}$ and $[\partial \Psi/\partial u]_k{}^{\#}$ accordingly. Then the inverse vector of the pseudo-control increment is Δy_{fin} and the corresponding increment of control will be determined by the system of the form

$$\begin{cases} \Delta y_{hi} = 2 \cdot \left[\frac{\partial f}{\partial y}\right]_h^{\#} \cdot \left(f(x_h,\ y_h,\ t_h) - \frac{x_h - x(t_i)}{\Delta t_i}\right), \\ \Delta u_{hi} = \left[\frac{\partial \Psi}{\partial u}\right]_h^{\#} \cdot \left(\Psi(y_h,\ u_h,\ t_h) - \frac{\Delta y_{hi}}{\Delta t_i}\right). \end{cases} \quad (6.60)$$

Forming of vectors $f(x_{\text{fin}}, y_{\text{fin}}, t_{fin})$ and φ $(y_{fin}, u_{\text{fin}}, t_{fin})$ is similar to the procedure for forming of vectors $f(x_i, y_i, t_i)$ and $\Psi(y_i, u_i, t_i)$. The set of their initial values consists of a set of vectors y_{fin} u_{fin}, known at a finite point in time t_{fin}. The change of the sign in the expression (6.60) takes into account the fact that the direction of differentiation changes from forward to reverse.

Let us consider in more detail the Eqs. (6.59) and (6.60). Expressions on their right side, assigned to a specific time interval Δt, are by difference operators, approximating the corresponding differential operators in the form

$$\begin{cases} Lx\left[\frac{\partial f}{\partial y}\right]^{\#}\cdot\left[\frac{dx}{dt}-f(x,\ y,\ t)\right]\cdot\Delta t^{-1}, \\ Ly\left[\frac{\partial\Psi}{\partial u}\right]^{\#}\cdot\left[\frac{dy}{dt}-\Psi(u,\ y,\ t)\right]\cdot\Delta t^{-1}. \end{cases} \quad (6.61)$$

The «~» symbol in this case is used to indicate the approximation. In the theory of difference schemes, there is a concept of right and left difference schemes. So, for example, for the differential operator $Lv = dv/dt,$ the corresponding difference operators on a two-point pattern ($t,t+\Delta$ and $t-\Delta,t$) can be written as [27]

$$Lv_{h+} = \frac{v(t+\Delta)-v(t)}{\Delta} \equiv v_{t+}; \quad (6.62)$$

$$Lv_{h-} = \frac{v(t)-v(t-\Delta)}{\Delta} \equiv v_{t-}, \quad (6.63)$$

where v_{t+}—right difference derivative;

v_{t-}—left difference derivative.

Linear combination of operators v_{t+} and v_{t-} allows you to form an infinite number of approximation schemes in the form

$$Lv_h(\sigma) = \sigma\cdot v_{t+} + (1-\sigma)\cdot v_{t-}, \quad (6.64)$$

where σ—any real number.

By settings $\sigma = 0.5$, we obtain the so-called central difference derivative [28].

In accordance with the above classification, the previously introduced expressions (6.59) form a system of right difference operators and the expressions (6.60) form a system of left difference operators, respectively. Linear combinations of the obtained increments with a coefficient $\boldsymbol{\sigma}$ form is an analog of the central difference derivative, for example, in the form

$$\Delta u_i\cdot\Delta t_c^{-1} = \sigma\cdot\Delta u_{\text{ifin}}\cdot\Delta t_c^{-1} + (1-\sigma)\cdot\Delta u_{\text{fini}}\cdot\Delta t_c^{-1} \quad (6.65)$$

In this case, we are talking about an analogy due to the fact that, in contrast to the standard form of formation of difference operators [28, 29], the expressions for Δu_{ifin} and Δu_{fini} are constructed according to a two-point pattern, not with internal neighbors, but with the extreme points of the segment $[t_{\text{in}},\ t_{\text{fin}}]$, which in the process of solving the boundary value problem form a convergent sequence.

If you integrate the system (6.38) simultaneously «on the right» and «on the left», respectively, with the controls

$$u_{i+1}^{+} = u_{\text{ifin}} + (\Delta u_{\text{ifin}}\cdot\Delta t_c^{-1})\Delta t_c; \quad (6.66)$$

$$u_{i+1}^{-} = u_{\text{fini}} + (\Delta u_{\text{fini}}\cdot\Delta t_c^{-1})\Delta t_c, \quad (6.67)$$

then a sequence of nodes t_{ifin} and t_{fini} will be formed, converging to some internal limit point of the segment [t_{in}, t_{fin}]. However, this method of calculating the controls is not always possible, since the «right end» of the trajectory is usually indeterminate for most problems, if not for all, then for some values of the phase coordinates and controls.

The problems that arise in the «counter» integration of the system (6.38) can be avoided by introducing a unidirectional («left to right») solution scheme. To do this, we assume that the current increment of the control at each step of integration will be determined by a linear combination (6.65) with the coefficient $\sigma = 0.5$, and the corresponding control in the next step will be determined by the equality of the form

$$u_{i+1} = 0.5 \cdot \left(u_i + u_{fin}\right) + \left(\Delta u_i \cdot \Delta t_c^{-1}\right) \cdot \Delta t_c, \tag{6.68}$$

or finally.

$$u_{i+1} = 0.5 \cdot (u_i + u_{\text{fin}}) + \Delta u_i. \tag{6.69}$$

As a result, we obtain finite sequences of nodes and points of the control function and trajectories, converging to the given boundary conditions on the right end of the segment [t_{in}, t_{fin}]. The condition for stopping the cycle by t_i, in this case, is to achieve equality t_i=t_{fin}.

The geometric meaning of the obtained scheme for determining the required increments of the vector of control functions can be illustrated by the example in Fig. 6.2.

From Fig. 6.2, it can be seen that the difference scheme used in the method forms such derivatives of the coordinates of the state, and consequently of the control, which provides movement from the current point of the trajectory to the final one along the shortest path. In this case, at each step of the integration of the dynamic system, the square of the current discrepancy of the coordinates of the state of the object of study is minimized.

Thus, the boundary value problem of the dynamics of the AC flight can be solved. However, in this book, it is assumed that the final solution must not only satisfy the given boundary conditions, but also deliver the optimal (in accordance with the accepted assumptions) value of the purpose function $W(x,u,t)$. Obviously, this is possible only if the basic solution obtained using the developed difference scheme (6.59) and (6.60) allows for variation with respect to the original function. The source of this variation in Eqs. (6.59) and (6.60) can be the function of intensity of the controls $h(t_i)$, in the choice of which there is some uncertainty. Exactly this uncertainty, makes it possible to optimize the initial solution from the point of view of the selected functional.

Therefore, it is obvious that the value of the increment of control at each step of the integration of the dynamic system (6.38) will largely be determined by the choice of the basic function of the intensity of the control $h(t_i)$ in (6.59) and (6.60).

By changing the value, and, if necessary, the sign of the argument of this function, the variation of required values of the corresponding derivatives of the state vector

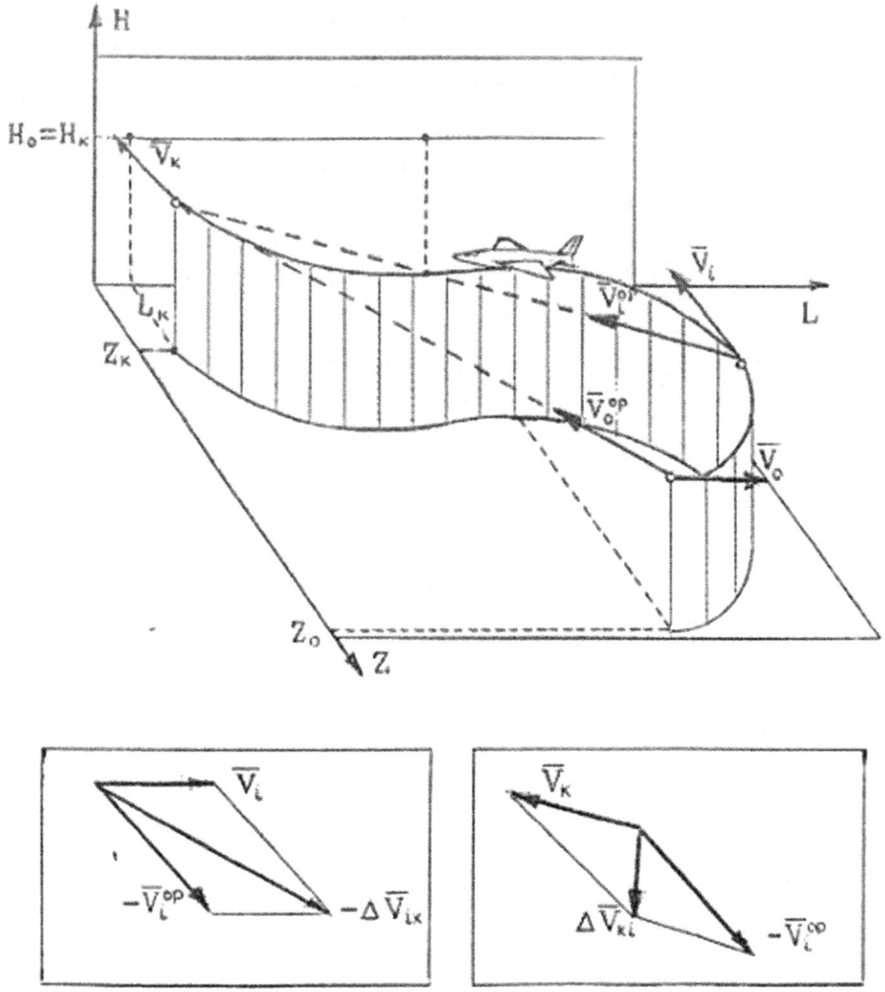

$$\Delta V_i = \sigma \Delta V_{iK} + (1-\sigma)\Delta V_{Ki}$$

Fig. 6.2 Geometric meaning of the difference scheme MDO

are realized, which, in accordance with the previously given proof, ensures, that the necessary optimality conditions are satisfied, taking into account the restriction on the nature of the trajectory's movement of the damaged AC.

6.3.3.3 Illustration of the Application of the Developed Difference Scheme to the Solution of the Simplest Boundary Value Problem. Determination of the Support Function of the Control Intensity

In the simplest case, when the behavior of the object of research is described, for example, by a dynamic system of the form (the case of a double integrating link)

$$\begin{cases} \dot{x} = y, \\ \dot{y} = u, \end{cases} \tag{6.70}$$

the scheme for calculating the increments of control can be written as a system of difference equations

$$\begin{cases} \Delta y_i = 2 \cdot \frac{(x_k - x_i)}{(t_k - t_i)} - \left(y_i + y_{fin}\right), \\ \Delta u_i = 0.5 \cdot \left(\frac{\Delta y_i}{h(t_i)} - u_i + u_{fin}\right). \end{cases} \tag{6.71}$$

Accordingly, the control at the (i+1)-th step of integration will be defined as the sum (6.69).

The system (6.71) was obtained as follows. According to the formula (6.65), the current control increment is represented as

$$\Delta u_i = 0.5 \cdot (\Delta u_{ifin} + \Delta u_{fini}). \tag{6.72}$$

Substituting the corresponding right sides for u_{ifin} and u_{fini} from (6.59) and (6.60) to (6.72), we get

$$\Delta u_i = 0.5 \cdot \left(\frac{(\Delta y_{ih} - \Delta y_{hi})}{h(t_i)} - u_i + u_{fin}\right), \tag{6.73}$$

where

$$\Delta y_{ih} = \frac{(x_{\text{fin}} - x_i)}{(t_{\text{fin}} - t_i)} - y_i; \tag{6.74}$$

$$\Delta y_{hi} = y_{\text{fin}} - \frac{(x_{\text{fin}} - x_i)}{(t_{\text{fin}} - t_i)}, \tag{6.75}$$

from where after the replacement $\Delta y_i = \Delta y_{ifin} - \Delta y_{fini}$, we get the system (6.71).

To solve the original boundary value problem by integrating the system (6.70) with the controls, found from (6.71), you need to set the function $h(t_i)$.

It follows from expressions (6.59) and (6.60) that the function $h(t_i)$ has the dimension of time. Obviously, by setting the function $h(t_i)$ in an arbitrary way $h(t_i)$ for $\Delta \mathbf{y}_i$,, known from the first equation of the system (6.71), you can get an arbitrary trajectory *x(t)* and the necessary control *u(t)* within a certain framework.

However, during the solving of this problem, an additional condition is set for compliance with the boundary conditions (6.39)–(6.44), which naturally limits the choice $h(t_i)$.

Consider the simplest scheme for selecting a basic grid function $h(t_i)$, which ensures the convergence of the solution of the difference scheme to the values of the original boundary value problem. For this purpose, we will use the expression introduced earlier for the square of the current residual in phase coordinates (6.45).

As before, we will look for the control $u(t)$ as a function that minimizes the current value of the residual $\rho(u)$ at each point of time $t_i \in [t_{in}, t_{fin}]$. As a result of performing operations, similar to (6.46)–(6.50), we get

$$u(t_i + \Delta t_c) = 2 \cdot ((x_{\text{fin}} - x(t_i)) \cdot \Delta t_i^{-1} - y(t)) \cdot \Delta t_i^{-1}, \tag{6.76}$$

where $u(t_i + \Delta t_c) \in$ locextr at each step of the problem to solve.

The expression, obtained in (6.76), is true both for integrating the system (6.70) «from left to right», and for the reverse case. In this case, the record form of the corresponding controls will take the form

$$u^+(t_{i+1}) = 2 \cdot ((x_{\text{fin}} - x(t_i)) \cdot \Delta t^{-1} - y(t_i)) \cdot \Delta t_i^{-1}; \tag{6.77}$$

$$-u^-(t_{i+1}) = 2 \cdot ((x_{\text{fin}} - x(t_i)) \cdot \Delta t^{-1} - y(t_{\text{fin}})) \cdot \Delta t_i^{-1}. \tag{6.78}$$

For the system record form (6.70), obtained in (6.71), it can be written that

$$\Delta u_i^- = u_{\text{fini}} - 2 \cdot (y_{\text{fin}} - (x_{\text{fin}} - x_i) \cdot \Delta t_i^{-1}) \cdot \Delta t_i^{-1}; \tag{6.79}$$

$$\Delta u_i^+ = 2 \cdot ((x_{\text{fin}} - x_i) \cdot \Delta t^{-1} - y_i) \cdot \Delta t_i^{-1} - u_{\text{ini}}, \tag{6.80}$$

from where we get

$$u_{i+1} = 0.5 \cdot (u_{\text{ini}} + u_{\text{fin}}) + 0.5 \cdot (\Delta u_i^+ + \Delta u_i^-), \tag{6.81}$$

or

$$\begin{aligned} u_{i+1} = {} & 0.5 \cdot (u_{ini} + u_{\text{fin}}) + 2 \cdot ((x_{\text{fin}} - x_i) \cdot \Delta t_i^{-1} - 0.5 \cdot (y_{\text{fin}} + y_i)) \cdot \Delta t_i^{-1} \\ & + 0.5 \cdot (u_{fin} - u_{ini}). \end{aligned} \tag{6.82}$$

Finally, we get

$$u_{i+1} = u_{\text{fin}} + 2 \cdot ((x_{fin} - x_i) \cdot \Delta t_i^{-1} - 0.5 \cdot (y_{fin} + y_i)) \cdot \Delta t_i^{-1}, \tag{6.83}$$

where from, when we compare with (6.73), it follows, that $h = 0.5 \cdot \Delta t_i$, where Δt_i is determined, according to (6.58).

It also follows from Eq. (6.83) that the tendency $y_i \rightarrow y_{\text{fin}}$ и $(x_{\text{fin}} - x_i) \cdot \Delta t^{-1} \rightarrow y_{\text{fin}}$ for $t_i \rightarrow t_{\text{fin}}$ automatically occurs in the case $u_i \rightarrow u_{\text{fin}}$, that is, the boundary conditions of the original problem are satisfied for all coordinates simultaneously.

Summarizing the above, we assume that the general form of the basic function of control's intensity can be represented as

$$h(t_i) = C \cdot (t_{\text{fin}} - t_i);\ C \neq 0, \quad -1 \leq C \leq 1, \tag{6.84}$$

where t_i change from t_{in} to t_{fin}; C—an arbitrary constant that characterizes the ratio of the real and «internal» time of the dynamic system for which the correction of control is supposed to be performed.

In problems of a higher level of complexity, the expression for the function of control's intensity $h(t_i)$ may differ significantly from the one presented above. In particular, we can assume that in the case of forming m controls in the space of n coordinates, this function will have the form of a diagonal matrix $n \times n$.

We note that setting the difference scheme in the form of (6.71) additionally allows us to determine the upper and lower bounds of the change of the function $h(t_i)$. To do this, you can use the known restrictions on the controls u_{max} and u_{min}. Successively substituting them in the left part of the Eq. (6.69) and taking into account (6.71), we get

$$(u_{\min} - u_{\text{fin}})/\Delta y_i \leq 0.5 \cdot h(t)^{-1} \leq (u_{\max} - u_{\text{fin}})/\Delta y_i. \tag{6.85}$$

Thus, in the case of varying the function $h(t)$, it is possible to narrow the range of variations, which facilitates the process of finding the extremums of the functionals.

6.3.3.4 Determination of Partial Derivatives of Functions of the Original Dynamic Aircraft Movement System

The example of a double integrating link is the most convenient for illustrating the operation of the method due to the fact that here the inverse matrices of partial derivatives are $[\partial f/\partial y]_i^{\#}$, $[\partial f/\partial y]_k^{\#}$, $[\partial \Psi/\partial u]_i^{\#}$, and $[\partial \Psi/\partial u]_k^{\#}$ take the form of a unit matrix E. The position of elements in the matrices of general form is determined as follows:

$$[\partial f/\partial y] = \begin{bmatrix} \frac{\partial f_1}{\partial y_1} & \frac{\partial f_1}{\partial y_2} & \cdots & \frac{\partial f_1}{\partial y_n} \\ \cdots & \cdots & \cdots & \cdots \\ \frac{\partial f_n}{\partial y_1} & \frac{\partial f_n}{\partial y_2} & \cdots & \frac{\partial f_n}{\partial y_n} \end{bmatrix}, \quad [\partial \Psi/\partial u] = \begin{bmatrix} \frac{\partial \Psi_1}{\partial u_1} & \frac{\partial \Psi_1}{\partial u_2} & \cdots & \frac{\partial \Psi_1}{\partial u_m} \\ \cdots & \cdots & \cdots & \cdots \\ \frac{\partial \Psi_n}{\partial u_1} & \frac{\partial \Psi_n}{\partial u_2} & \cdots & \frac{\partial \Psi_n}{\partial u_m} \end{bmatrix}. \tag{6.86}$$

Returning to the original problem, we note that in this case, when finding the increments Δy_i and Δu_i, it is necessary to calculate the inverse matrices of the partial derivatives $[\partial f/\partial y]_i^{\#}$, $[\partial f/\partial y]_{fin}^{\#}$, $[\partial \Psi/\partial u]_i^{\#}$, and $[\partial \Psi/\partial u]_{fin}^{\#}$, which are obtained using the pseudo-conversion operation [22] from the corresponding matrices of partial derivatives of vector functions f and Ψ by arguments y and u.

Taking into account the introduced notation, the corresponding components of the matrices of partial derivatives will have the following form:

(a) partial derivatives with respect to the coordinates of the pseudo-control vector

$$\frac{\partial f_1}{\partial y_1} = \cos\theta \cdot \cos\phi; \tag{6.87}$$

$$\frac{\partial f_1}{\partial y_2} = -V \cdot \sin\theta \cdot \cos\phi; \tag{6.88}$$

$$\frac{\partial f_1}{\partial y_3} = -V \cdot \cos\theta \cdot \sin\phi; \tag{6.89}$$

$$\frac{\partial f_2}{\partial y_1} = \sin\theta; \tag{6.90}$$

$$\frac{\partial f_2}{\partial y_2} = V \cdot \cos\theta; \tag{6.91}$$

$$\frac{\partial f_2}{\partial y_3} = 0; \tag{6.92}$$

$$\frac{\partial f_3}{\partial y_1} = -V \cdot \cos\theta \cdot \sin\phi; \tag{6.93}$$

$$\frac{\partial f_3}{\partial y_2} = V \cdot \sin\theta \cdot \cos\phi; \tag{6.94}$$

$$\frac{\partial f_3}{\partial y_3} = -V \cdot \cos\theta \cdot \cos\phi; \tag{6.95}$$

б) partial derivatives with respect to the coordinates of the control vector

$$\frac{\partial \Psi_1}{\partial u_1} = \frac{\cos(\alpha - \phi^*)}{m}; \tag{6.96}$$

$$\frac{\partial \Psi_1}{\partial u_2} = -\frac{P \cdot \sin(\alpha - \phi^*) + \partial X_f / \partial\alpha}{m}; \tag{6.97}$$

$$\frac{\partial \Psi_1}{\partial u_3} = 0; \tag{6.98}$$

$$\frac{\partial \Psi_2}{\partial u_1} = \frac{\sin(\alpha - \phi^*) \cdot \cos\gamma}{m \cdot V}; \tag{6.99}$$

$$\frac{\partial \Psi_2}{\partial u_2} = -\frac{\left(P \cdot \cos(\alpha - \phi^*) + \partial Y_f/\partial\alpha\right) \cdot \cos\gamma}{m \cdot V}; \quad (6.100)$$

$$\frac{\partial \Psi_2}{\partial u_3} = -\frac{\left(P \cdot \sin(\alpha - \phi^*) + Y_f\right) \cdot \sin\gamma}{m \cdot V}; \quad (6.101)$$

$$\frac{\partial \Psi_3}{\partial u_1} = -\frac{sin(\alpha - \phi^*) \cdot \sin\gamma}{m \cdot V \cdot \cos\theta}; \quad (6.102)$$

$$\frac{\partial \Psi_3}{\partial u_2} = -\frac{\left(P \cdot \cos(\alpha - \phi^*) + \partial Y_f/\partial\alpha\right) \cdot \sin\gamma}{m \cdot V \cdot \cos\theta}; \quad (6.103)$$

$$\frac{\partial \Psi_3}{\partial u_3} = -\frac{\left(P \cdot \sin(\alpha - \phi^*) + Y_f\right) \cdot \cos\gamma}{m \cdot V \cdot \cos\theta}. \quad (6.104)$$

All elements of the matrices (6.86) are determined using the boundary conditions of the boundary value problem, which are recalculated with each cycle of controls's forming for the left end of the trajectory and once for the right end of the trajectory.

The resulting system of difference Eqs. (6.59) and (6.60) is closed and allows to obtain unambiguous values of the coordinates of the control vector $u(t_i)$. The type and nature of changes of the controls over time are determined by the previously introduced basic function $h(t_i)$. The nature of the function $h(t_i)$ is selected from the condition of satisfying the boundary conditions at the right end. In addition, it will be shown that we can obtain an extreme solution from the point of view of some optimality criterion by the variation of the functional dependence $h(t_i)$.

The found extreme solution will be the only one in the given class of functions that define the form $h(t_i)$.

6.3.3.5 Determination of the General Sequence of Application of the Developed Difference Scheme for Calculating the Aircraft Trajectory

Consider the sequence of operations performed to form the current controls and the corresponding trajectories:

1. Using the values of the boundary conditions of the boundary value problem (6.39)–(6.44) and Eqs. (6.59), (6.60), we find successively the vectors of increments of pseudo-controls and controls corresponding to the form of the function $h(t_i)$. Due to the nonlinearity of the system (6.38), matrices $[\partial f/\partial y]_i^{\#}$and $[\partial \Psi/\partial u]_i^{\#}$ are calculated and inversed on each cycle of control's formation, and the matrices $[\partial f/\partial y]_k^{\#}[\partial \Psi/\partial u]_k^{\#}$ are calculated and inversed once.
2. Summing up Δu_i with the original controls: $u_{i+1} = 0.5 \cdot \left(u_{ini} + u_{fin}\right) + \Delta u_i$.
3. Check the resulting control to satisfy the constraints of the inequality type (if they are given). If they exceed the corresponding limit values, they are recalculated or hard re-assigned to the type u=u_{max}.

4. We enter the received control into the model (6.38).
5. We integrate the values of the phase coordinates during the cycle time Δt_c. If there are restrictions of the inequality type on the parameters of the phase coordinates and their derivatives, then a check is made for their satisfaction. If the set limit values are exceeded, the correction of the control is realized to ensure movement along the border of the permissible area. The algorithm of correction is discussed in the next section.
6. Memorize the new values of the vectors x_i, y_i, as well as the new control u_i and time t_i.
7. Check the condition $t_{\text{fin}} - t_i > 0$: if it is satisfied, the cycle repeats with c item 1; otherwise, the calculation process is terminated, and the parameters of the trajectory and the corresponding controls are stored in the memory of the calculator.

The variational function $h(t_i)$, corresponding to the obtained acceptable solution can be approximated by some bounded piecewise continuous curve of a given smoothness class. The obtained new function is the initial one for the variational problem of finding the extremum of a given functional W. The mechanism for solving this problem, built on the basis of direct variational methods, as well as using the ideas of the theory of synthesis of optimal control [1, 2], is the content of item 6.3.5.

Certain results of the functioning of the MDO algorithms are presented in works [20, 30–37].

This algorithm allows you to build a set of real-time realizable trajectories from the current or any other point of space or give a negative result «There are no realizable trajectories».

6.3.4 *Organization of the Procedure for Maintaining the Restrictions on Coordinates of State and Control Vectors*

Since the structure of the method for solving boundary value problems of flight dynamics, described in item 6.3.3, provides automatic satisfaction of all boundary conditions at the ends of a given integration section, we consider the procedure for taking into account the constraints on the phase coordinates and control functions within the segment $[t_{in}, t_{fin}]$.

In general case, there are two ways to set constraints. First, it is the formation of the outer boundaries of the range of acceptable values of phase coordinates and controls. Second, it is the assignment of some local closed manifolds within a given domain. In both cases, it is necessary to provide conditions for «non-piercing» of the boundary hypersurface.

Let us consider separately the mechanism for taking into account the constraints on the functions of control and on the phase coordinates of an arbitrary AC, when it moves in the atmosphere.

Assuming the parameters of the control functions to be non-inertial values, we will use the method of movement along a given boundary in order to take in account the restrictions on their current values: when the corresponding coordinate of the vector *u(t)* reaches the upper or lower boundary, its further growth or decrease stops. At the same time, the remaining coordinates that are within the acceptable values, retain complete freedom.

Since the object of study considered in this book is a complex technical system that is subjected to significant physical exertion during flight, it is necessary to ensure such conditions for its functioning that would guarantee the successful solution of the tasks assigned to it even in conditions of stricter restrictions as a result of failures. It is obvious that a decrease in the maneuverability of the AC leads to a decrease in the space of its permissible controls and, accordingly, requires the imposition of stricter restrictions on the parameters of its phase coordinates.

Consider the procedures for maintaining the constraints. Let's write the expressions (6.59), (6.65), and (6.69) in the following form:

$$\begin{aligned}\Delta y_{ifin} &= 2\left(\frac{\partial f}{\partial y^T}\right)^{\#}\left(\frac{x_{fin}-x_i}{t_k-t_i} - f(x_i,\ y_i,\ t_i)\right);\\ \Delta u_{ifin} &= \left(\frac{\partial \Psi}{\partial u^T}\right)^{\#}\left(\frac{\Delta y_{i\,fin}}{h(t_i)} - \Psi(y_i,\ u_i,\ t_i)\right);\\ \Delta u_i &= \left(\sigma\cdot\Delta u_{i\,fin} + (1-\sigma)\cdot\Delta u_{fini}\right);\\ u_{i+1} &= (\Delta u_i + u_i)\ .\end{aligned}$$

$$\begin{aligned}\Delta y_{ifin} &= 2\left(\frac{\partial f}{\partial y^T}\right)^{\#}\left(\frac{x_{fin}-x_i}{t_{fin}-t_i} - f(x_i,\ y_i,\ t_i)\right);\\ \Delta u_{ifin} &= \left(\frac{\partial \Psi}{\partial u^T}\right)^{\#}\left(\frac{\Delta y_{ifin}}{h(t_i)} - \Psi(y_i,\ u_i,\ t_i)\right);\\ \Delta u_i &= \left(\sigma\cdot\Delta u_{ifin} + (1-\sigma)\cdot\Delta u_{fini}\right);\\ u_{i+1} &= (\Delta u_i + u_i).\end{aligned} \tag{6.105}$$

Here is the symbol "⌐" means the operation of limiting the upper and lower levels of the corresponding signals.

As an example, we will write the restriction under the number

$$\Delta y_{\text{ifin}} = \begin{cases}\Delta y_{\text{ifin}}^{\max},\ if\ \Delta y_{\text{ifin}} > \Delta y_{\text{ifin}}^{\max},\\ \Delta y_{\text{ifin}},\ if\ \Delta y_{\text{ifin}}^{\min} \le \Delta y_{\text{ifin}} \le \Delta y_{\text{ifin}}^{\max},\\ \Delta y_{\text{ifin}}^{\min},\ if\ \Delta y_{\text{ifin}} < \Delta y_{\text{ifin}}^{\min}\end{cases} \tag{6.106}$$

where $\Delta y_{\text{ifin}}^{\max} = y_{\text{ifin}}^{\max} - y_{\text{ifin}}, \Delta y_{\text{ifin}}^{\min} = y_{\text{ifin}}^{\min} - y_{\text{ifin}}$.

Here and further, we mean a component-by-component restriction of the component of vectors *x*, *y*, *u*, and their derivatives.

Restriction 1 allows us to prevent the exit from the operational area of the coordinates of the velocity vector $\bar{V} = [V \quad \theta \quad \varphi]^T$.

Restriction 2 provides a limit on the rate of change of derivatives $\dot{L}$, $\dot{H}$, $\dot{Z}$if it is required. In addition, by zeroing the corresponding component, they provide a constraint on this coordinate (*L, H, Z*).

Restriction 3 provides a constraint on derivatives $\dot{V}$, $\dot{\theta}$, and$\dot{\phi}$, if necessary.

When implementing constraints 1–3, the coefficient σ in (6.105) is equal to one accordingly. This ensures that the requirements of satisfaction of the final conditions do not affect the process of maintaining one or the other restrictions.

Restriction 4 provides a cut-off of the unrealizable value of the signals of control in the same way (6.106).

Restriction 5 provides a taking into account the dynamic properties of the executive circuit. In this case, it is the diagonal matrix of the corresponding transfer functions. In other cases, it may be the transfer functions of the executive devices of the control system.

The generated control signal on the i-th cycle, taking into account all possible constraints, enters the model (6.38), and it is integrated during the cycle time. In this case, the corresponding section of the trajectory is calculated.

Thus, in contrast to direct variational methods, the use of the object's model allows you to immediately take into account and implement component-by-component constraints on the state vectors, controls, and their derivatives.

The most difficult problem is the problem of maintaining multidimensional constraints in the form of various kinds of hypersurfaces described by equations of the form

$$\varphi(x) = 0, x \in R^{\mathrm{n}}. \tag{6.107}$$

The process of maintaining such restrictions can be divided into 3 stages:

- ensuring a smooth «landing» of the trajectory on the constraint surface (CS) without its piercing;
- movement on the CS, taking into account its configuration $\varphi(L, H, Z) = 0$;
- going off the CS, when it does not interfere with the propagation of the trajectory in the direction of the final conditions.

Successive implementation of the conditions for entering on the CS, moving along it, and going off, obviously, will allow you to get an acceptable basic solution.

First stage

The implementation of the first stage is based on a continuous analysis of the ratio of two time intervals.

The first time interval represents the time of achievement T_{ach} CS in the case that the subsequent movement can be predicted based on the history of the movement.

The second time interval is the available time T_{av}, for which, using some given control, allocated to the constraint task U_{constr}, you can cancel the rate of approximation with the CS to zero.

When the equality of these two times $T_{\text{ach}} = T_{av}$ is reached, the constraint mode is turned on and the object's trajectory tangentially falls on the CS.

We will reveal the formal component of the described approach.

There is a current point of the trajectory $x = [L \quad H \quad Z]^T$. There is some prehistory of the movement using which you can determine the equation of the trajectory $T(L, H, Z) = 0$.

Next, the joint solution of the equations of the surface and the trajectory is carried out

$$\begin{cases} \phi(L,\ H,\ Z) = 0, \\ T(L,\ H,\ Z) = 0 \end{cases} \tag{6.108}$$

in order to find the meeting point of the trajectory with the surface x_n=$[L\ H\ Z]_n{}^{\text{T}}$.

Knowing x and x_n, you can calculate the distance of the current point to the surface

$$d = [(x_n - x)^{\text{T}} \cdot (x_n - x)]^{0.5} \tag{6.109}$$

Differentiating this expression in time, we get

$$\dot{d} = -\frac{\dot{x}^T (x. - x)}{d}. \tag{6.110}$$

Using the last two expressions, we get an expression for calculating the time to reach the surface

$$T_{ach} = -\frac{d}{\dot{d}} = -\frac{d^2}{\dot{x}^T (x_n - x)}. \tag{6.111}$$

where $\dot{x}$ corresponds to the object's Eqs. (6.31)–(6.33).

Similarly (6.111), in the derivative space by an order of magnitude higher, the available time T_{av} is determined

$$T_{av} = -\frac{\dot{d}}{\ddot{d}_{av}}, \tag{6.112}$$

where $\ddot{d}_{av}$—the available value of the second derivative of the distance to CS. It is defined as (basically) u_{constr}, so and its own properties of the control object.

Expression to calculate $\ddot{d}$ has the following form:

$$\ddot{d} = -\frac{[\ddot{x}^T (x_n - x) - \dot{x}^{\text{T}} \dot{x}] \cdot d + [\dot{x}^{\text{T}} (x_n - x)]^2 \cdot d^{-1}}{d^2}, \tag{6.113}$$

where $\ddot{x} = \begin{bmatrix}\ddot{L} & \ddot{H} & \ddot{Z}\end{bmatrix}^T$, and the corresponding expressions allow one to calculate the values of the second derivatives

$$\begin{aligned}
\ddot{L} &= \dot{V}\cos\theta\cdot\cos\phi - V\sin\theta\cdot\cos\phi\cdot\dot{\theta} - V\cos\theta\cdot\cos\phi\cdot\dot{\phi};\\
\ddot{H} &= \dot{V}\sin\theta + V\cos\theta\cdot\dot{\theta};\\
\ddot{Z} &= -\dot{V}\cos\theta\cdot\sin\phi + V\sin\theta\cdot\sin\phi\cdot\dot{\theta} - V\cos\theta\cdot\cos\phi\cdot\dot{\phi}.
\end{aligned} \quad (6.114)$$

In the elements of this expression $\dot{V}$, $\dot{\theta}$, $\dot{\varphi}$, defined by Eqs. (6.28)–(6.30), it is necessary to substitute the values of the components of the vector of control $u_{\text{constr}} = \left[P\ \alpha\ \gamma\right]^T_{\text{constr}}$.

Equating the expressions (6.111) and (6.112), we get the distance, at which, with the selected control u_{constr}, it is necessary to implement the mode of maintaining restrictions

$$-\frac{d}{\dot{d}} = -\frac{\dot{d}}{\ddot{d}_{act}}, \quad (6.115)$$

from which

$$d_{\text{constr}} = \frac{\dot{d}^2}{\ddot{d}_{av}} \quad (6.116)$$

Thus

$$u = \begin{cases} u_{cr.giv.}, \ if\ d > d_{\text{constr}}, \\ u_{\text{constr.}}, \ if\ d \le d_{\text{constr}}. \end{cases} \quad (6.117)$$

With a smooth «landing» of the trajectory on the CS ($d = 0$, $\dot{d} = 0$), the first stage ends.

Second stage

It consists of the formation of such control u_{constr} which provides the formation of the trajectory of movement along CS in the direction of finite conditions.

In order for the trajectory of movement of the AC not to pierce CS, it is necessary that at the moment of its contact with the boundary of the area defined by the equation $\varphi(L, H, Z) = 0$, and then in the course of movement, the vector of the derived coordinates of the state $\dot{x}$ was directed tangentially to it. An illustration of this condition is shown in Fig. 6.3. This condition can be provided by introducing the following matrices of partial derivative $[\partial X_{\text{bound}}/\partial x]^T$ and $[\partial Y_{\text{bound}}/\partial y]^T$, which are defined as:

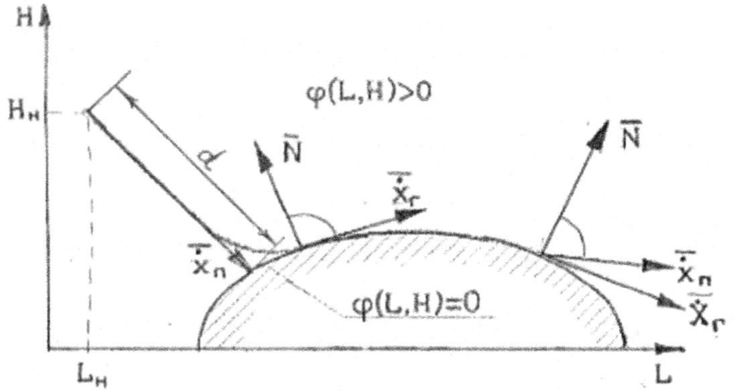

Fig. 6.3 Diagram of the trajectory's reaching to the surface boundary $\varphi(L, H) = 0$, movement along it, and the fulfillment of the condition of the going off

$$\left[\frac{\partial X_{bound}}{\partial x}\right]^T = \begin{bmatrix} \frac{\partial L_{bound}}{\partial L} & \frac{\partial L_{bound}}{\partial H} & \frac{\partial L_{bound}}{\partial Z} \\ \frac{\partial H_{bound}}{\partial L} & \frac{\partial H_{bound}}{\partial H} & \frac{\partial H_{bound}}{\partial Z} \\ \frac{\partial Z_{bound}}{\partial L} & \frac{\partial Z_{bound}}{\partial H} & \frac{\partial Z_{bound}}{\partial Z} \end{bmatrix}; \tag{6.118}$$

$$\left[\frac{\partial Y_{bound}}{\partial y}\right]^T = \begin{bmatrix} \frac{\partial V_{bound}}{\partial L} & \frac{\partial V_{bound}}{\partial H} & \frac{\partial V_{bound}}{\partial Z} \\ \frac{\partial \theta_{bound}}{\partial L} & \frac{\partial \theta_{bound}}{\partial H} & \frac{\partial \theta_{bound}}{\partial Z} \\ \frac{\partial \phi_{bound}}{\partial L} & \frac{\partial \phi_{bound}}{\partial H} & \frac{\partial \phi_{bound}}{\partial Z} \end{bmatrix} \tag{6.119}$$

The introduction of matrices (6.118) and (6.119) is due to the following considerations.

Let there be some surface on the path of the object of study from the starting point to the end point that constraints its movement (Fig. 6.3). Setting the form CS in the form $\varphi(L, H, Z) = 0$, we can control the fact of its piercing by the trajectory of the movement of the object of study. In this case, as shown in Fig. 6.3, function $\varphi(L, H, Z)$ will change the sign from «+» to «−».

When the trajectory reach (Fig. 6.3) on CS $\varphi(L, H, Z) = 0$, then the movement of a dynamic object must be implemented by the tangent to this surface. The direction of the tangent is determined by the normal $N = [\partial\varphi/\partial L, \partial\varphi/\partial H, \partial\varphi/\partial Z]^T$.

Presented in item 6.3.3.2, the difference scheme for calculating the increments of control functions allows to determine the vector of derived state coordinates at each point in three-dimensional space including on the surface $\varphi(L,H,Z)=0$

$$\dot{R}_i = \left[\dot{L}_i,\ \dot{H}_i,\ \dot{Z}_i\right]^T. \tag{6.120}$$

The theory of mathematical analysis follows that the movement along a tangent to a given surface is determined by the equality of the form

$$N^T \cdot \dot{R} = 0. \tag{6.121}$$

Revealing this equality, we obtain the following equation:

$$[\partial\phi/\partial L] \times \dot{L} + [\partial\phi/\partial H] \times \dot{H} + [\partial\phi/\partial Z] \times \dot{Z} = 0, \tag{6.122}$$

from which follows, that for the continuation of the movement on the surface with the derivatives of the vector $\dot{R}_i$ necessary to express in terms of the components of the Eq. (6.122) one or two from it components. The movement in this direction is forced and is determined by the corresponding derivative, for example

$$\dot{Z}_{forc.} = -[\partial\phi/\partial Z_{bound.}]^{-1} \times \left([\partial\phi/\partial L] \times \dot{L} + [\partial\phi/\partial H] \times \dot{H}\right), \tag{6.123}$$

or

$$\dot{Z}_{forc.} = -\left([\partial Z_{bound.}/\partial L] \times \dot{L} + [\partial Z_{bound.}/\partial H] \times \dot{H}\right) \tag{6.124}$$

So the matrices of partial derivatives (6.118) and (6.119) are formed. Accordingly, the difference equations for determining the increments of pseudo- controls and control functions take the form

$$\begin{cases} \Delta y_{ifin} = 2 \cdot \left[\dfrac{\partial f}{\partial y}\right]_i^{\#} \cdot \left(\left[\dfrac{\partial X_{bound}}{\partial x}\right]^T \cdot \left[x_{fin} - x(t_i)\right]\Delta t_1^{-1} - f(x_i, y_i, t_i)\right), \\ \Delta u_{ifin} = \left[\dfrac{\partial \psi}{\partial u}\right]_i^{\#} \cdot \left(\left[\dfrac{\partial Y_{bound}}{\partial y}\right]^T \cdot \Delta y_{ifin} \cdot h(t_i)^{-1} - \psi(y_i, u_i, t_i)\right) \end{cases} \tag{6.125}$$

Defining apriori the components of the matrices of partial derivatives (6.118), (6.119) and implementing formulas of the type (6.125), we obtain a strict condition for observance of the given restrictions. The limiting case for such an arrangement of the solution will be the case when all the phase coordinates have reached their limits.

If, moving along the boundary of the area of acceptable values, it is impossible to fulfill the boundary conditions of the original problem, then the resulting trajectory is considered unrealizable and cannot be used as a basic solution.

Third stage

It provides current control of the possibility of going off the CS in the case when it ceases to prevent the free propagation of the trajectory in the direction of the final conditions of the boundary value problem.

The criterion of going off is the value of the angle between the normal to CS at the current point and the vector of the required values of the derivatives for solving the boundary value problem. As soon as this angle becomes less than 90 degrees, the CS envelope mode stops, and then the trajectory is constructed in free space. The formal expression of the criterion for going off has the form

$$\xi = \arccos \frac{N^T \cdot \dot{x}_n}{\left\| N^T \right\| \cdot \|\dot{x}_n\|} < 90^\circ. \tag{6.126}$$

The schematic diagram of the implementation of the proposed algorithm for maintaining constraints on phase coordinates has the form shown in Fig. 6.4.

Next, in Figs. 6.5 and 6.6, the results of the practical application of the developed algorithm directly for the problems of spatial motion of the AC under the conditions of restrictions on the coordinates of the state are presented.

For an illustration of the adaptive properties of this approach to maintaining constraints, we present the conclusions of the laws of limiting the angle of attack and the minimum acceptable flight altitude. They are reduced to the form that is used in traditional ACS, the laws of which are given for comparison.

Limiting the angle of attack

Time of achievement α_{ach} is determined by the formula

$$T_{ach} = \frac{\alpha_{ach} - \alpha}{\dot{\alpha}} \tag{6.127}$$

The available time to decrease the rate of increase of the angle of attack is determined by the expression of the form

$$T_{av} = \frac{0 - \dot{\alpha}}{\ddot{\alpha}} \tag{6.128}$$

Equating these two times and revealing the expression $\ddot{\alpha}$, we get the angle of attack of the activation of the system for limits of the angle of attack (SLAA)

$$\alpha_{constr} = \alpha_{ach} + \frac{\dot{\alpha}^2}{\left(\bar{M}_z^{\omega_z} + \bar{M}_z^{\dot{\alpha}} - \bar{Y}^{\alpha}\right) \cdot \dot{\alpha}\left(\bar{M}_z^{\omega_z} \cdot \bar{Y}^{\alpha} + \bar{M}_z^{\dot{\alpha}}\right) \cdot \dot{\alpha} + \bar{M}_z^{\delta_e} \cdot \delta_e^{SLAA}}. \tag{6.129}$$

For comparison, here is a model law SLAA

$$\alpha_{constr} = \alpha_{ach} - K\dot{\alpha} \tag{6.130}$$

Even a cursory analysis (6.129) speaks in favor of the proposed approach which responds flexibly to change $\alpha_{constr.}$, when changing the damping characteristics, AC

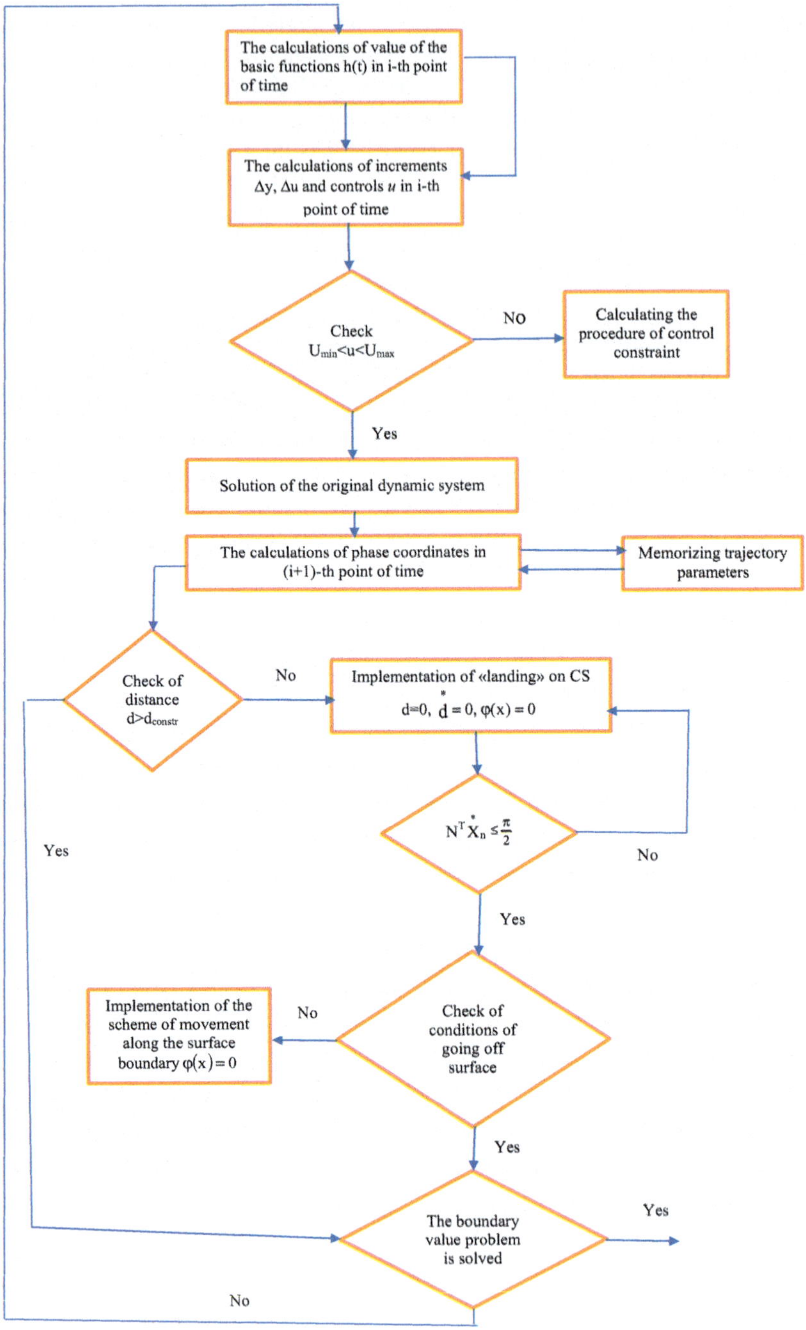

Fig. 6.4 Block diagram of the algorithm for maintaining constraints

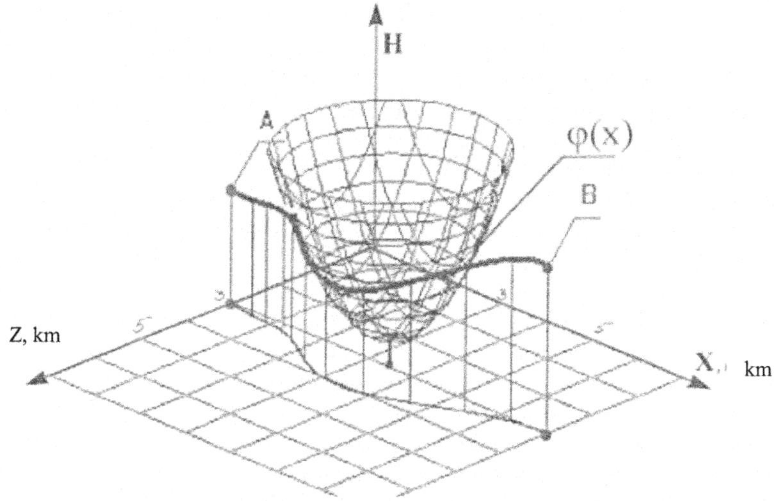

Fig. 6.5 The results of the practical application of the developed algorithm directly for the problems of spatial motion of the AC in the conditions of restrictions on the coordinates of the state

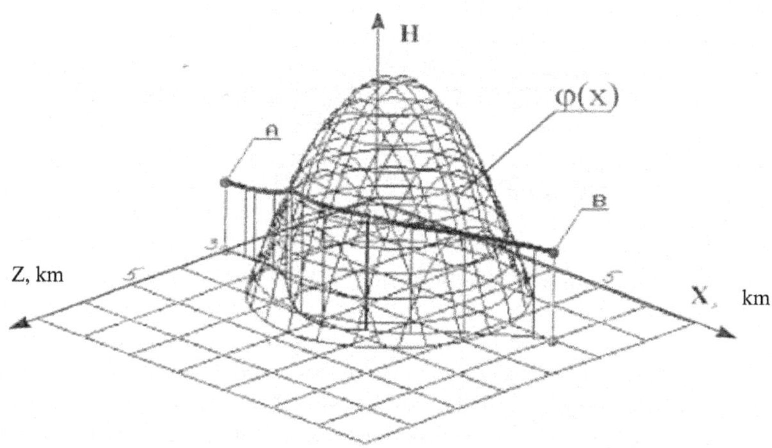

Fig. 6.6 The results of the practical application of the developed algorithm directly for the problems of spatial motion of the AC in the conditions of restrictions on the coordinates of the state

stability, and efficiency of the elevator wheel, as opposed to the traditional approach (6.130).

Minimum height limit (the mode «Withdrawal from dangerous height»).

Let's write down the values of the corresponding time intervals

$$T_{ach} = \frac{H_{\min} - H}{\dot{H}}, \; T_{av} = -\frac{\dot{H}}{\ddot{H}}. \tag{6.131}$$

Equating these two expressions by choosing a fixed overload value $n_y^{withdrawal}$ as the control and expanding the expression $\ddot{H}$, as a result, we get

$$H_{constr} = H_{\min} + \frac{\dot{H}^2}{g \cdot \left(n_x \sin\theta + n_y^{withdrawal} \cos\gamma \cdot \cos\theta - 1\right)} \tag{6.132}$$

The law of mode «Withdrawal» of the traditional ACS has the form

$$H_{Constr} = H_{\min} - K\dot{H} \tag{6.133}$$

It can be seen that, in this case, the proposed approach to maintaining the limit of height has undeniable advantages from the point of view of adaptation and full consideration of the spatial position of the AC during the descent.

6.3.5 *The Procedure for Optimizing the Basic Trajectory Movement of Aircraft*

As already noted in item 6.3.1 of this section, the final stage of forming the optimal solution of the task of control is the variation of the acceptable basic trajectory of a dynamic object in order to achieve an extreme value of a given functional on it.

In works [19, 38], it is shown that the ordered variation of the parameters of the basic function of the state coordinate vector allows us to obtain a trajectory that provides an extremum to the functional and satisfies the conditions of the original boundary value problem. Such a trajectory is called optimal.

Let us consider from these positions the possibilities of variation of the basic solution obtained using the difference scheme developed in item 6.3.3. In principle, there are two possible approaches to solving this problem.

The first involves the use of a well-known strategy of direct variational methods which consists of directly varying the current parameters of the coordinates of the state vector of a dynamic object. The second approach can be built on the use of the internal capabilities of the developed difference scheme and assumes the presence of a certain function, the change of which leads to a change in the derived parameters of the movement of a dynamic object.

The use of the methodology of direct variational methods, as already noted at the beginning of this section, is associated with significant difficulties regarding the feasibility of the resulting trajectory, as well as compliance with restrictions on the parameters of the phase coordinates. Therefore, we will proceed directly with the consideration of the second version for optimizing the basic solution.

Earlier, in item 6.3.3.2, in the course of the conclusion of the main expressions of the difference schemes (6.59) and (6.60), it was already noted that their application allows for some uncertainty in the formation of the basic trajectory. This uncertainty is due to the initial form of the setting of function of the control intensity $h(t_i)$ introduced in the consideration. At the same time, it should be noted that any form of representation of this function, that does not lead to its values going beyond the area bounded by the expression (6.85), does not affect the convergence of the solution to the values of the boundary conditions at the ends of the segment $[t_{\text{in}}, t_{\text{fin}}]$. This, in turn, means that any change $h(t_i)$ within the boundaries of inequality of type (6.85), causing a change in the derivatives of the coordinate vector of the basic trajectory at the same time does not violate its acceptable character.

The presence of this feature suggests that the considered function of the intensity of controls $h(t_i)$, in addition to its main part in forming a basic acceptable solution to the original boundary value problem, can be used to vary the resulting trajectory in order to optimize it, taking into account the restriction of the type of trajectory in accordance with item 6.3.3.2.

Let's consider the nature of the influence of the function $h(t_i)$ on the view of the trajectory of a dynamic object at the simplest example of the problem of optimal control for a double integrating link. In this case, as already shown earlier, the form of the function $h(t_i)$ is determined by the value of coefficient C before the converging difference $t_{\text{fin}}-t_i$. To illustrate the physical meaning of this coefficient, we can use the formula (6.71), in which the ratio $\Delta y_i \cdot h(t_i)^{-1}$ corresponds to the required value of the second derivative of the state coordinate vector—d^2x_n/dt^2. Accordingly, selecting from the expression $\Delta y_i \cdot h(t_i)^{-1}$ ratio $\Delta y_i \cdot C^{-1}$, one can see that it is the equivalent of the required increment of the function of the first derivative of the state coordinate vector—$d(dx_n/dt)$. The value of the coefficient C in this case acts as a control parameter of the intensity of the vector's change Δy_i : the higher the value of C, the smaller the required increment of the derivative dx_n/dt, and on the contrary.

In the difference scheme under consideration, the change in coefficient C and the change in function $h(t_i)$ are equivalent. Thus, it turns out that the variation of the values of the basic function $h(t_i)$ at each moment of time t_i is equivalent to the variation of the values of the derivatives of the required vector of coordinates of the state of the dynamic object $d(dx_n/dt)$ at the moments t_i.

Earlier, when proving the Euler equation in item 6.3.2 in terms of the developed method, it was shown that the variation of the current derivative (without taking into account the limitations of item 6.3.3.2) allows us to achieve results in optimizing the trajectory of a dynamic object that are completely identical to those obtained by directly changing the vector of state coordinate. And, if a double integrating link is sufficient to select the optimal value C once (as shown in item 6.3.3.4), then in more complex cases, it is obvious to assume that each time interval $[t_i, t_{i+1}]$ corresponds to its optimal value of the coefficient C_i.

The problem of determining the optimal values of the coefficient C_i for each segment $[t_i, t_{i+1}]$ is equivalent to the problem of finding the optimal form of the basic grid function $h(t_i)$.

We will search for the basic function $h(t_i)$ in numerical form. To do this, we divide the given integration segment $[t_{\text{in}}, t_{\text{fin}}]$ into p parts and fix the values of the basic function $h(t_i)$, defined by (6.94), at the points of the partition. As a result, we get p basic points of the function $h(t_i)$, varying which we will look for an extreme solution in the sense of the given functional. The form of the functional W, in this case, will be determined by the values of the p parameters of the original basic function $h(t_i)$, i=1,2,…,p.

In fact, we got an implicit function of p variables $W(h_1, h_2, \ldots, h_p)$, defined on the interval $[t_{\text{in}}, t_{\text{fin}}]$. It is assumed that the set of all possible values of the functional W is non-empty and finite, so we can talk about the existence of some best function $W_o(h_1, h_2, \ldots, h_p)=W_{\text{extr}}$ on it, which corresponds to one or more trajectories and controls called optimal ones.

We will assume that the resulting implicit function $W(h_1, h_2, \ldots, h_p)$ is twice differentiable by variables $h_1, h_2, \ldots, h_p$ on the set of all valid values h_i. In this case, according to the Taylor expansion theorem [21], it can be written that in the neighborhood of a point $\alpha = 0$, functional $W(h)$ has the form

$$W(h + \alpha \cdot d) = W(h) + \alpha \times g(h)^T \times d + 0.5 \times \alpha^2 \times d^T \times G(h) \times d + O(\alpha^3), \tag{6.134}$$

where α—step of variation of the corresponding variables;

h—some point in the coordinate space $h_1, h_2, \ldots, h_p$;

d—direction in p-dimensional space;

$g(h)$—gradient of a given purpose function $W(h)$, equal by definition

$$g(h) \equiv [\partial W/\partial h_1, \ldots, \partial W/\partial h_p]^T; \tag{6.135}$$

$G(h)$—a Hesse matrix, composed of the second derivatives of the original functional with respect to variables h_i

$$G(h) \equiv \begin{bmatrix} \frac{\partial^2 W}{\partial h_1^2} & \cdots & \frac{\partial^2 W}{\partial h_1 \partial h_p} \\ \cdots & \cdots & \cdots \\ \frac{\partial^2 W}{\partial h_1 \partial h_p} & \cdots & \frac{\partial^2 W}{\partial h_p^2} \end{bmatrix}. \tag{6.136}$$

Here, the rate of change of the function $W(h)$ during movement along the direction d from the point h is given by $g(h)^{\mathrm{T}} \cdot d$, and the curvature $W(h)$ along d is determined by the product $d^{\mathrm{T}} \cdot G(h) \cdot d$. The error of approximation by the formula (6.134) has, according to [13, 21], the order $\|\alpha\|^3$.

To search for an extreme value of the given purpose function W_{extr}, we use the principle of unconditional minimization (maximization) [21], according to which the algorithm for determining the optimal set of variable parameters (in this case h_1, $h_2, \ldots, h_p$) requires that the value of the functional W decreases (increases) at each iteration, i.e. the condition of descent (ascent) is honored, which consists of the fact that:

$$W_{k+1} < W_k, (W_{k+1} > W_k) \text{ for all } k \geq 0. \tag{6.137}$$

In this case, the sufficient condition of the extremum, when satisfied, the calculations are stopped, can be formulated as follows:

$$\text{a)} \|g(\text{h}_{extr})\| = 0; \tag{6.138}$$

$$\text{b) matrix} G(h_{extr}) \text{ is determined positively} \tag{6.139}$$

The model scheme for solving the obtained p-dimensional problem of unconditional optimization, according to [21], can be represented as a certain fixed sequence of operations:

1. Check the conditions of compliance with the stop (6.138) and (6.139). If the conditions are honored, then the calculations are stopped and the current solution h_{Tfin} is taken as the desired one. Otherwise, we proceed to the next step.
2. Calculation of the non-zero p-dimensional vector d_{fin} of the direction's search.
3. The calculation of step length α_{fin}, ensuring inequality $W(h_{\text{fin}} + \alpha_{\text{fin}} \cdot d_{\text{fin}}) < W(h_{\text{fin}})$.
4. Recalculation of the decision evaluation. We believe $h_k + \alpha_k \cdot d_k \rightarrow h_{k+1}$, $k \rightarrow k + 1$ and return to item 1.

In order to ensure the convergence of the model scheme 1–4 with a new order of representation of the function W, an additional condition is introduced that relates to its set of level [21]: for a given function W and some number β of sets, $L(\beta) = \{h : W(h) \leq \beta\}$ must be limited and closed. In this case, algorithm 1–4 generates a sequence of points for which

$$\lim_{k \to \infty} g(h_{mk}) = 0. \tag{6.140}$$

Convergence of this type, according to [21], is usually called global. The estimation of the proximity of the obtained numerical optimal solution to its possible analytical analog, according to the theorem of convergence of the coordinate descent method [21], is determined by the inequalities:

$$|\text{u}_{\text{num.}} - \text{u}_{\text{extr}}| \leq \text{L} \cdot \text{H}/\text{l}; \tag{6.141}$$

$$0 \leq |\text{W}_{\text{num.}} - \text{W}_{\text{extr}}| \leq \text{L}^{2} \cdot \text{H}^2/(2 \cdot 1), \tag{6.142}$$

where $u_{\text{num.}}$, $u_{\text{extr,}}$ and $W_{\text{num.}}$, W_{extr}—numerical and analytical optimal controls and values of functionals, respectively;

L and l—constants that limit the area of changes in the values of the second derivatives of the Hesse matrix, i.e.

$$1 \leq \|G(h)\| \leq L; \tag{6.143}$$

$$H = \left[\sum_{i=1}^{n} \Delta h_i^2 \right]^{1/2}. \tag{6.144}$$

Here the norm of the matrix $G(h)$ is defined as the maximum of the modulus of the partial derivative $\partial W/(\partial h_i \cdot \partial h_j)$ for all i and j, where $i = 1,...,p$ and $j = 1,...,p$.

Thus, as a result of successive variation by all points of a given basic function $h(t_i)$, satisfying the conditions of the original boundary value problem, we obtain the following set h_1, h_2, …, h_p, which ensures the fulfillment of the conditions (6.138), (6.139) and, accordingly, the extreme value of the functional W.

The variation procedure itself can be performed in two ways. By convention, they can be called a method of «continuous» search and a search by scheme «window». An illustration of both approaches is given in Figs. 6.7 and 6.8.

From Fig. 6.7, it can be seen that in the case of the implementation of the «continuous» scheme, the variation of the basic solution in each section of the segment $[t_{in}, t_{fin}]$ assumes the recalculation of all subsequent parameters of the trajectory. The implementation of this calculation scheme allows us to find a solution to the optimal control problem with the minimum amount of occupied RAM memory. However, the time required for calculations is expected to be significantly higher than, for example, in the case of using the principle of «window» of organization of calculations. In the latter case, as seen in Fig. 6.8, the variation and recalculation of the trajectory parameters are carried out only in the i-th section of the basic trajectory. In this case, the coordinates and control functions in all other sections of the given integration interval $[t_{in}, t_{fin}]$ must be stored in the RAM memory of the computing device, which

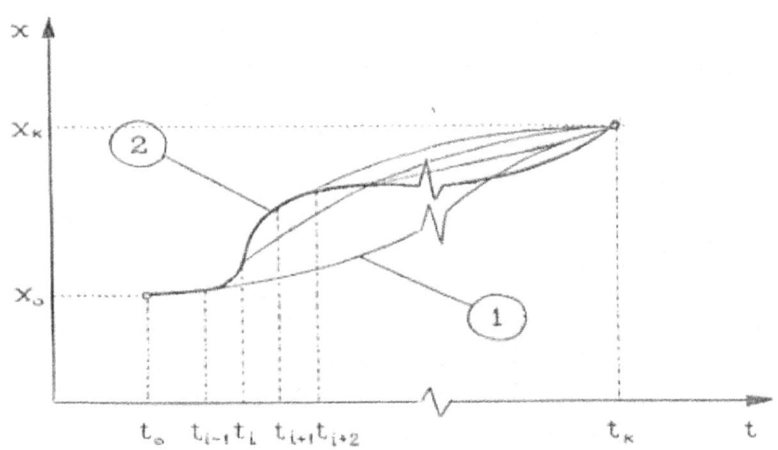

1 – the basic solution 2 – optimal solution

Fig. 6.7 Scheme of «continuous» search for optimal solution

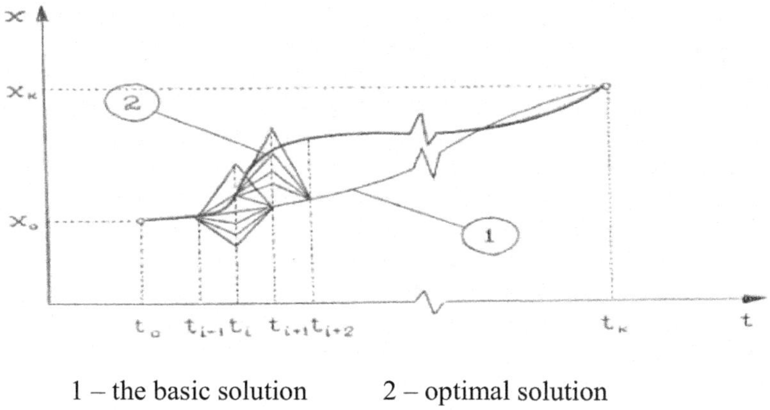

Fig. 6.8 Search for optimal solution by scheme «window»

requires significant amounts of RAM memory and indicts high requirements for the speed of information circulation inside the computer [29, 30].

Obviously, both approaches lead to the same solutions and the choice of a particular calculation scheme is completely determined by the specific capabilities of the researcher.

In conclusion, we note that, in this case, the meaning of optimizing the trajectory of the AC movement does not generally correspond to the classical understanding of *min W*. The set of possible trajectories is narrowed by the scheme of setting the derivatives and their variations determined by minimizing the norm (6.45).

6.4 Estimation of the Reliability of the Method of Direct Optimization Using Examples of Solving the Simplest Optimal Control Problems

Since the proposed method of direct optimization in principle does not depend on the type of the original system of differential equations and the nature of the desired functional, then in order to assess the reliability of the solutions obtained as a result of its application, we will conduct a comparative analysis of the solutions of the basic simplest tasks of the classical theory of the calculus of variations, allowing, in addition to numerical, analytical representation [31].

As the main indicators of the adequacy of the obtained numerical solutions to the analytical ones, we take such parameters as the value of the given functional on the desired extremal, the standard deviation of the parameters of the numerical and analytical functions of the state and control coordinates throughout the entire integration segment, as well as the relative values of the residuals at its ends. It is assumed, that a numerical solution, that provides, within the given accuracy of the calculations ∈, the satisfaction of the boundary conditions of the original problem and the inequality for functionals of the form

$$|W_{extr} - W_{num.}| \leq e_1, \tag{6.145}$$

$$\sum_{j=1}^{f} \|\Delta x_j\| \leq \varepsilon_2, \tag{6.146}$$

where $f = (t_{\text{fin}} - t_{\text{in}})/\Delta t_c$;

Δx_j—discrepancy between the current numerical and optimal solutions;

$$\|x_k - x(t_k)\| \leq \varepsilon_3, \tag{6.147}$$

will be considered absolutely optimal.

In this case, we will take the value of the *Wextr* functional as basic, obtained on the basis of indirect variational methods (classical calculus of variations or the principle of maximum of L. Pontryagin) [4, 5, 8, 11].

To conduct a comparative analysis, we will use the examples of solving the main types of extreme problems, encountered in the theory of classical calculus of variations (CCV) given in [4, 5, 11].

Example 1. As the first, we consider the simplest problem of optimal control by a linear object, described by an equation of the form

$$x = \dot{u}, \text{ where } u \in U = [0, 1][\text{TSF1}]\ for\ \forall t \in [0, 1] \tag{6.148}$$

with initial conditions: $x(0) = 0$, $x(1) = 3/4$.

The functional in this problem is given in the integral form and has the form

$$W(x(\cdot), u(\cdot)) = \int_0^1 (u^2 - 4x)dt \to inf. \tag{6.149}$$

Application of the theory CCV and the principle of maximum of L. Pontryagin gives the only analytical solution that delivers the absolute minimum to the functional (6.149) under the given initial conditions.

In accordance with the MDO principles presented above, we write the function of the increment of control for this task

$$\Delta u_i = 0.5 \cdot ((x_{\text{fin}} - x_i)h(t_i)^{-1} - u_i + u_{\text{fin}}), \tag{6.150}$$

where $u_{i+1} = 0.5 \cdot (u_i + u_{fin}) + \Delta u_i$.

$x_{fin} = x(1)$—the value of the x coordinate at a finite time t_{fin};

x_i—the value of the x coordinate at a current point of time t_i ($x_{in} = x(0)$);

$h(t_i)$—the function of intensity of control;

i—index of the current moment of time in the case of integration by t from 0 to 1.

Integration of Eq. (6.148) on the segment [0,1] with the control u_i, obtained from (6.150), when $u(0) = 1$, u(1) = 0, and $h(t_i)$=0.5· ($t_{fin}-t_i$) gives the values of the trajectory $x(t_i)$ at each point of time t_i.

The resulting numerical solution and the results of the application of the analytical approach can be represented graphically in the form of a function of control $u_{extr}(t_i)$ and trajectory $x_{extr}(t_i)$, corresponding to this control of the trajectory [32]. As seen in Fig. 6.9a, the graphs of the numerical and analytical solutions coincide (the standard deviation (6.146) of the parameters of the state and control coordinate functions does not exceed 1% of their maximum value). The calculated values of the functionals differ by no more than the value of the calculation error

$$W(x(\cdot), u(\cdot))_{extr} = -1.17665000; \tag{6.151}$$

$$W(x(\cdot), u(\cdot))_{num.} = -1.17645025. \tag{6.152}$$

The Euclidean norm of residuals (6.147) at the x coordinate is zero. Therefore, according to the previously introduced extremality criterion, we can assume that the received control and the corresponding trajectory are the solution to the original task.

Example 2 Consider a more complex example of the problem of optimal control. Let the object of research be described by a dynamic system

$$\begin{cases} \dot{x} = y \\ \dot{y} = u, \quad u \in U = \{u : |u| \le 1\}, \ \forall t \in [0, 1]. \end{cases} \tag{6.153}$$

The boundary conditions have the form: $x(0) = 0$, $x(1) = 0$.

The purpose function is presented by an integral of the form

$$W(x(\cdot)) = \int_0^1 x(t)dt \to inf. \tag{6.154}$$

In [11], it was proved that the solution to this problem exists and is unique.

We apply the method of direct optimization (MDO) and the system obtained on its basis (6.71) to the solution of this problem. In the system (6.71), there is a designation for the function of the speed of movement of the object of study along the trajectory $x(t) = y$. Since the boundary values of the derivatives are not given in the original task, we define the set of known initial and final conditions with the values y (0) = $-1/2$ and y(1) = 1/2, which are determined from the condition of the extremum of the functional (6.154).

Solving successively systems (6.71) and (6.153), when $u(0)$=1, $u(1) = 1$, and $h(t_i)$=0.5·($t_{fin}-t_i$), we get the desired function of control and the trajectory of the movement of object in the interval [0,1] [33].

Graphs of the corresponding functions, obtained by numerical and analytical methods, are shown in Fig. 6.9b. As you can see from the figure, both methods give the same solutions. The calculated deviations (6.146) and (6.147) are almost zero,

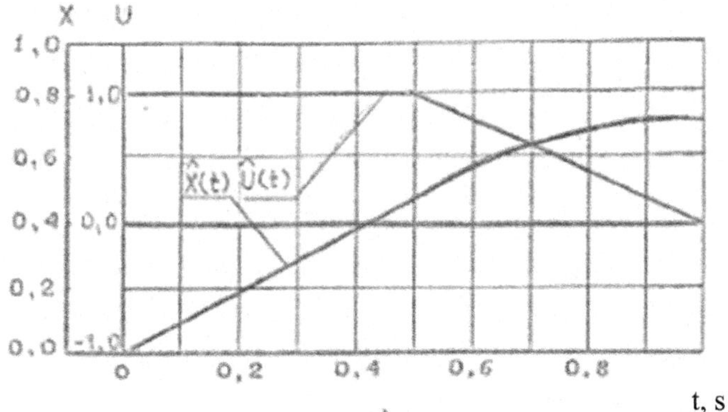

a)

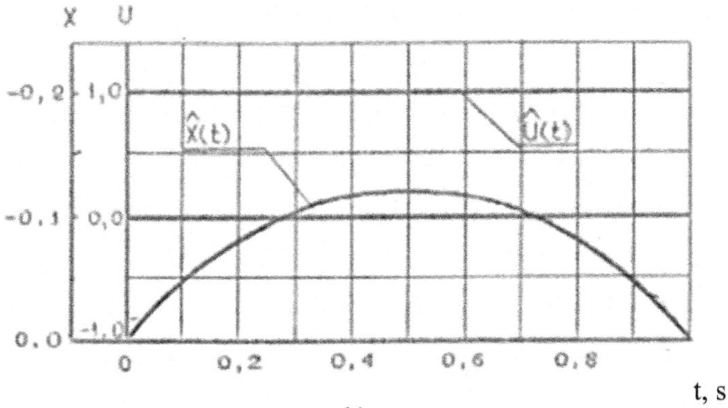

b)

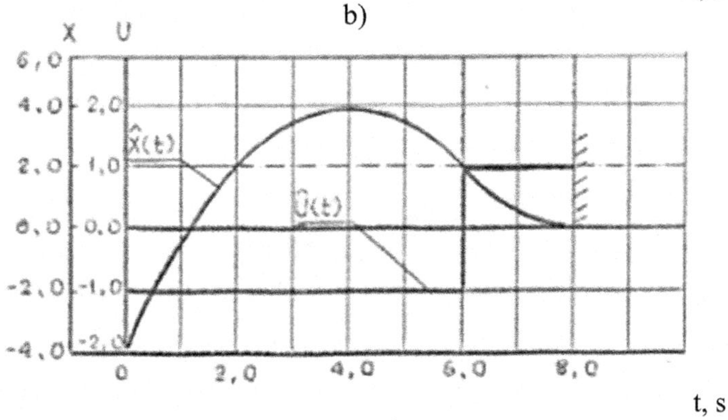

c)

Fig. 6.9 Examples of solving the simplest tasks of optimal control using the method of direct optimization

and the values of the functionals differ by no more than the value of the machine accuracy of the operations [34]

$$W(x(\cdot))_{extr} = -0.08333201, \tag{6.155}$$

$$W(x(\cdot))_{num.} = -0.08332930. \tag{6.156}$$

Obviously, both solutions are solutions to the original problem.

Example 3 Consider a problem of operating speed. Let be given a dynamical system (6.153), where $u \in U = \{u : |u| \leq 1\}$, $\forall t \in [0, \ t_k]$, and the given boundary conditions have the form

$$x(0) = -4, x(t_{fin}) = 0, \ y(0) = 4, \ y(t_{fin}) = 0. \tag{6.157}$$

The functional in this problem is given as a minimum value of the time of the end of the movement of the material point

$$W(x(\cdot), t_{fin}) = t_{fin} \to inf. \tag{6.158}$$

It follows from the CCV theory and the principle of maximum [8, 11] that the solution to the problem exists and is unique.

This problem does not require further determination by the phase coordinates and their first derivative, since the set of boundary conditions is complete, with the exception of the values of the control functions at the points of time t_{in} and t_{fin}. In general, these values are included in the number of variable parameters and are sought from the condition for ensuring the extremum of the functional W.

The physical meaning of the original problem and the assumption that there is a control switch in the course of movement allow us immediately to select the initial and final controls in the form $u(0) = -1, \ u(t_{fin}) = 1$.

In accordance with the general approach of the method of direct optimization, the form of the function of the increment of control for this problem is determined by the system (6.71).

Successively compressing the time interval, we get that $t_{fin} = 8$. Further, solving systems (6.71) and (6.153), when $t_{\text{fin}} = 8, \ u(0) = -1, \ u(t_{\text{fin}}) = 1$, and $h(t_i) = 0.5 \cdot (t_{\text{fin}} - t_i)$, we get the control and trajectory that almost coincide with the analytical dependencies given earlier. The graphs of the numerical and classical solutions are shown in Fig. 6.9c. The calculated residuals (6.146) and (6.147) are still practically zero, and the values of the functionals in this problem coincide with definition [35].

Since the condition of equality of the numerical and analytical functionals for an acceptable value of the residual is satisfied, we will talk about the adequacy of the two presented approaches to solving the original problem.

The examples, discussed above, were used to demonstrate the adequacy of the solutions obtained using the MDO and CCV methods: the same initial values of the boundary conditions and constraints were matched with the functions of control and the corresponding trajectories that were adequate in their nature and value of the functional. At the same time, the process of finding extreme solutions was simplified by further definition of a set of boundary conditions for the tasks, set in accordance with the requirements of the method described earlier [36].

To analyze the behavior of the method under consideration in conditions of complete certainty of the original problem, we consider an example of solving the Lagrange problem in the following statement.

Example 4 The dynamic system, describing the state of an object, is defined as

$$\begin{cases} \dot{x} = y, \\ \dot{y} = u - x, \quad u \in U =]-\infty, +\infty[, \quad \forall t \in \left[0, \ \frac{\pi}{2}\right], \end{cases} \tag{6.159}$$

where $x(0) = 1$, $x(\pi/2) = 0$, $y(0) = 1$, $y(\pi/2) = -1 - \pi/2$, $u(0) = -4/\pi$, $u(\pi/2) = 0$.

The functional is presented in the integral form

$$W(x(\cdot), \ u(\cdot)) = \int_0^{\pi/2} u^2(t)dt \rightarrow in \ f. \tag{6.160}$$

The original problem can be solved using the CCV methods and the principle of maximum by L.S. Pontryagin.

The application of the MDO gives the following scheme for determining the increments of the control function:

$$\begin{cases} \Delta y_i = 2 \cdot (x_k - x_i) \cdot (t_k - t_i)^{-1} - (y_i + y_k), \\ \Delta u_i = 0.5 \cdot (\Delta y_i \cdot h(t_i)^{-1} - u_i - x_i + u_k + x_k \end{cases}. \tag{6.161}$$

Setting the initial and final controls in the form $u(0) = 0$, $u(1) = 0$, and assuming $h(t_i) = 0.38 \cdot (t_k - t_i)$, as a result of the successive solution of systems (6.161) and (6.159), we obtain the numerical values of the quasi-optimal control and the trajectory of the movement of the object on the segment $[0, \pi/2]$. Graphs of numerical and analytical solutions are presented in Fig. 6.10. From Fig. 6.10, it can be seen that the resulting numerical solution does not completely coincide with the function of optimal control $u_{opt}(t)$.

Since this approximation was performed without using the variation of the intensity function $h(t_i)$, then we can assume that there is an opportunity to improve the original solution. Therefore, assuming $h(t_i)$ as a polynomial constructed from the nodes of the function $h(t_i)=0.38 \cdot (t_k - t_i)$, where $p = 1.0.14$, and using the methodological approach described above, we obtain a function $p = 14$ variables: $W(h_1,..,.h_p)$ instead of the functional (6.160). Solving task $W(h_1, \ldots, h_p) \rightarrow inf$ by the method of coordinate descent, we obtain a new form of the intensity function $h(t_i)$ and the corresponding optimal control and trajectory of the object. The new numerical and

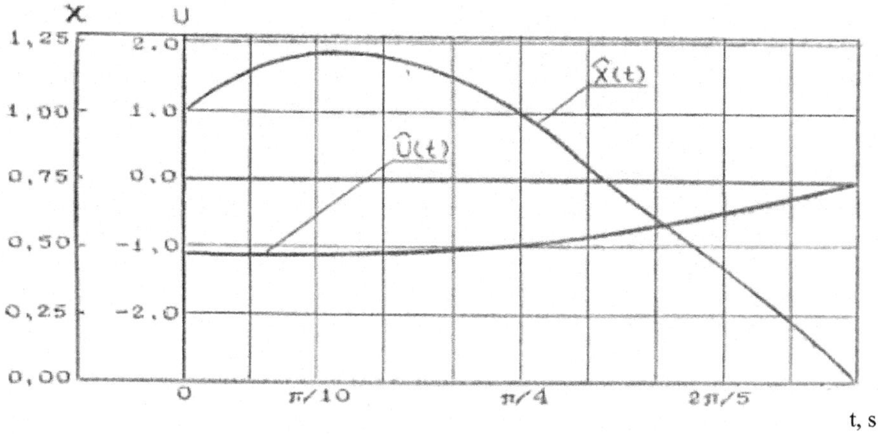

Fig. 6.10 Solution of the simplest problem of optimal control (version #4)

analytical solutions are almost identical as evidenced by the values of the residuals (6.146) and (6.147), which are close to zero. In this case, the corresponding values of the functionals are equal to

$$W(x(\cdot),\ u(\cdot))_{extr} = 1.27005645; \tag{6.162}$$

$$W(h_1, \ldots, h_p)_{num.} = 1.27045217. \tag{6.163}$$

Thus, the condition of equality of the numerical and analytical functionals at the valid value of the residual is fulfilled, which allows us to conclude that the obtained solution is adequate to its basic prototype.

The above examples of the tasks of optimal control and their solutions provided for the presence of at least one non-zero boundary condition either for the value of the phase coordinate or for the value of its derivative. However, this requirement is not mandatory. This can be confirmed by the solution of the task of a boundary value problem in which both the initial and final conditions are zero.

Example 5. Consider a dynamical system, that describes the state of an object with equations of the form

$$\begin{cases} \dot{x} = y, \\ \dot{y} = u,\ u \in U =]-\infty,\ +\infty[,\ \forall t \in [0, 1], \end{cases} \tag{6.164}$$

where $x(0) = 0$, $x(1) = 0$, $y(0) = 0$, $y(1) = 0$.

For this system, you need to find such controls and their corresponding trajectories that would deliver a minimum to functional

$$I(x(\cdot),\ u(\cdot)) = \int_0^1 (u^2(t) - 48 \cdot x(t))dt \to i\ n\ f. \tag{6.165}$$

The solution to the problem (6.164, 6.165) can be obtained by classical methods [8, 11].

Using the standard MDO scheme, set by the system (6.71), in the case u(0) = $u(1) = 2$, we obtain the numerical function of control *u(t)* and the corresponding trajectory *x(t)*, the graphs of which are shown in Fig. 6.11. As can be seen from the graphs and the results of calculating the formulas (6.146), and (6.147)and, the numerical and analytical solutions of the original problem coincide along the entire length of the segment *[0,1]*. The corresponding values of the functionals are equal to

$$W(x(), u())_{extr} = -0.79991111; \tag{6.166}$$

$$W(h_1, \ldots, h_p)_{num.} = -0.79625766 \tag{6.167}$$

Again, we have the equality of the numerical and analytical values of the functionals with the coincidence of the current trajectories and controls and the acceptable value of the residual. Obviously, both solutions can be considered adequate.

The analysis of the reliability of the method of direct optimization will be incomplete if we ignore the solution of the problems of optimal control of real dynamical systems, the behavior of which is described by a system of nonlinear differential equations under the conditions of constraints.

A classic example of this kind of problem is the brachistochrone problem for which it is customary to use the CCV or the principle of maximum by L.S. Pontryagin [2, 5]. Since a detailed description of the solution in this section would take up too much

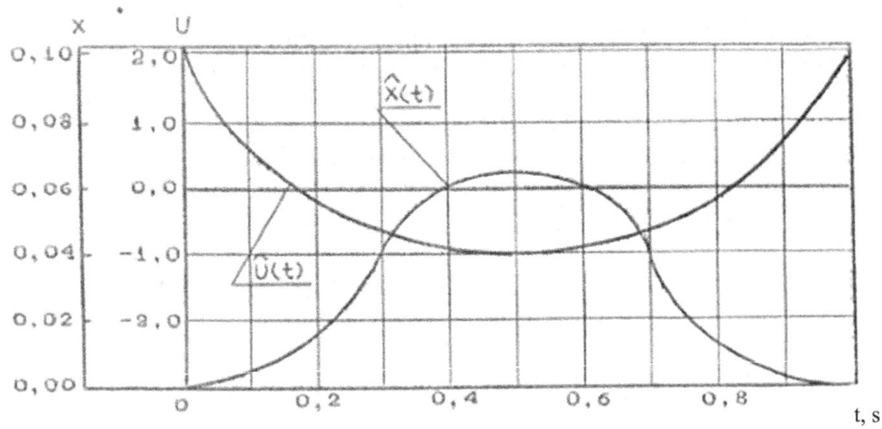

Fig. 6.11 Solving the simplest problem of optimal control (version #4)

space, we only note that the application of the method of direct optimization to the brachistochrone problem with fixed boundary conditions allows us to obtain solutions that are absolutely identical to the classical version of the solution. A version of one of them is shown in Fig. 6.12.

Examples of solving more complex problems can be the problems of assessing the remaining maneuvering capabilities of the AC after damage in modes, characterized by the maximum possible values of controls across all channels. The need for this type of mode is obvious in the case of performing maneuvers in the shortest possible time.

A typical and simplest example of the problem of evaluating the remaining capabilities of the AC after damage is the case of flat acceleration with the AC turning in the horizontal plane under conditions of restrictions on the functions of control [37].

In this case, it was assumed that the behavior of a dynamic object is described by a standard system of differential equations for the movement of the AC

$$\begin{bmatrix} \dot{X} \\ \dot{H} \\ \dot{Z} \end{bmatrix} = \begin{bmatrix} V \cdot \cos\theta \cdot \cos\varphi \\ V \cdot \sin\theta \\ V \cdot \cos\theta \cdot \sin\varphi \end{bmatrix}, \tag{6.168}$$

$$\begin{bmatrix} \dot{V} \\ \dot{\theta} \\ \dot{\varphi} \end{bmatrix} = \begin{bmatrix} g \cdot (n_x - \sin\theta) \\ g \cdot (n_x \cdot \cos\gamma - \cos\theta) \cdot V \\ -g \cdot n_y \cdot \sin\gamma \cdot [V \cdot \cos\theta]^{-1} \end{bmatrix}. \tag{6.169}$$

At the same time, conditional constraints of the inequality type were imposed on the roll and longitudinal overload functions. In practice, they are due generally not only to operational values, but also to the impact of damage or failure.

The requirement to ensure a minimum time of turnaround, set in the task, determines the obvious choice of the turning trajectory with the maximum acceptable roll angle. Assuming, that the modulus of the roll angle γ does not exceed 0.5 rad, we can obtain the only possible trajectories of the optimal turn.

It should be noted that the entire process of obtaining a solution to the problem of optimal control fits into the following scheme:

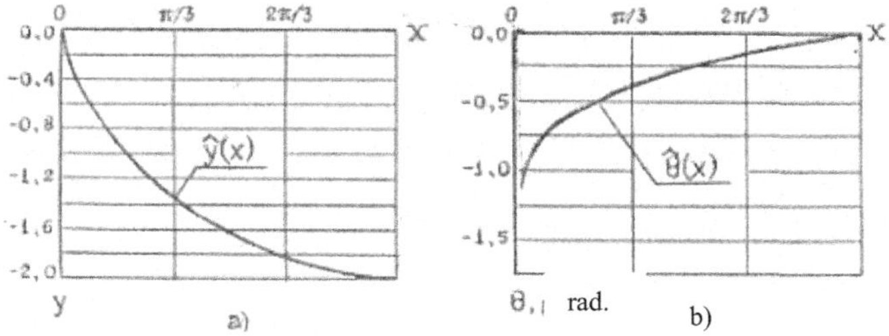

Boundary conditions

	H	K
x	0.0	π
-y	0.0	2.0
θ	0.0	0.0
V	0.0	π

Fig. 6.12 Results of solving the «brachistochrone» problem by the method of direct optimization

In a closed convex domain of acceptable values of the function of derivative intensity, bounded by inequalities, we are looking for a form of the basic function *h(t)* that would ensure the transfer of the initial dynamical system from a given initial state to a given final state in compliance with the current restrictions on the phase coordinates.

For conditions of item 1, the function *h(t)* is modified so that the final solution delivers an extreme value to the given functional *W* of the original task.

The resulting solution is proposed to be considered optimal in a given class of curves and the accepted scheme for the formation and variation of functions of the vector of derivatives. To solve the problems of control of damaged AC, this solution seems to be the most engineering-based.

6.5 Synthesis of the Laws of Control of the Pilot Circuit Based on the Method of Direct Optimization

The versatility of the MDO can be demonstrated by the example of solving problems of controlling a damaged AC in the pilot circuit. The control system is built, according to the methodology described above. By analogy with the trajectory contour, which was discussed in detail in the previous items, we will decompose the system of differential equations describing the movement in the pilot circuit, and build a control system based on the principles of the MDO.

The system of equations, describing the movement of the AC, is presented in the form [39]

$$\begin{aligned}\dot{\alpha} =& \omega_z - tg\ \beta(\omega_x \cos\ \alpha - \omega_y \sin\ \alpha) - \frac{g}{V\ \cos\ \beta}\left[(n_x \sin\ \alpha + n_y \cos\ \alpha) - \right.\\ &\left. - (\sin\ \vartheta\ \sin\ \alpha + \cos\ \vartheta\ \cos\ \gamma\ \cos\ \alpha)\ ;\right.\\ \dot{\beta} =& (\omega_x \sin\ \alpha + \omega_y \cos\ \alpha) - \frac{g}{V}\{[(n_x \cos\alpha - n_y \sin\ \alpha)\ \sin\ \beta - n_z \cos\ \beta] - \\ &- [(\sin\ \vartheta\ \cos\ \alpha - \cos\ \vartheta\ \cos\ \gamma\ \sin\ \alpha) \sin\ \beta + \cos\ \vartheta\ \sin\ \gamma\ \cos\ \beta]\ \};\\ \dot{\gamma} =& \omega_x - tg\ \vartheta(\omega_y \cos\ \gamma - \omega_z \sin\ \gamma);\end{aligned} \tag{6.170}$$

$$\begin{aligned}\dot{\omega}_x &= \frac{I_{yy}}{I_{xx}} M_x + \frac{I_{xy}}{I_{xx}} M_y + \frac{I_{yy}(I_{yy} - I_{zz}) + I_{xy}^2}{I_{xx}} \omega_y\omega_z - \frac{I_{xy}(I_{xx} + I_{yy} - I_{zz})}{I_{xx}} \omega_x\omega_z\ ;\\ \dot{\omega}_y &= \frac{I_{xx}}{I_{yy}} M_y + \frac{I_{xy}}{I_{yy}} M_x + \frac{I_{xx}(I_{xx} - I_{zz}) - I_{xy}^2}{I_{yy}} \omega_x\omega_z + \frac{I_{xy}(I_{xx} + I_{yy} - I_{zz})}{I_{yy}} \omega_y\omega_z\ ;\\ \dot{\omega}_z &= \frac{1}{I_{zz}} M_z - \frac{I_{yy} - I_{xx}}{I_{zz}} \omega_x\omega_y + \frac{I_{xy}}{I_{zz}}(\omega_x^2 - \omega_y^2)\ ;\end{aligned}$$

Let's imagine the equations, included in the pilot circuit (6.170), as a two-level hierarchy of systems of equations. The first level of the hierarchy is represented by a system of equations of the form

$$\dot{\mu} = \Psi(\mu\ ,\ \eta), \tag{6.171}$$

where $\mu = \left[\alpha,\ \beta,\ \gamma\right]^T$—state vector of the system;
$\eta = \left[\omega_z,\ \omega_y,\ \omega_x\right]^T$—vector of pseudo-control.
After the Taylor series expansion, the system (6.171) will take the form

$$\dot{\mu}_k = \Psi(\mu_{k-1}\ ,\ \eta_{k-1}) + \left.\frac{\partial\ \Psi}{\partial\ \mu^{\mathrm{T}}}\right|_{k-1} \Delta\ \mu_k + \left.\frac{\partial\ \Psi}{\partial\ \eta^{\mathrm{T}}}\right|_{k-1} \Delta\ \eta_k. \tag{6.172}$$

Hence, the increment of the pseudo-control vector at each step is determined by the following expression:

$$\Delta\ \eta_k = \left.\frac{\partial\ \Psi}{\partial\ \eta^{\mathrm{T}}}\right|_{k-1}^{-1} \left(\dot{\mu}_k - \Psi(\mu_{k-1}\ ,\ \eta_{k-1}) - \left.\frac{\partial\ \Psi}{\partial\ \mu^T}\right|_{k-1} \Delta\ \mu_k\right), \tag{6.173}$$

where $\dot{\mu}_k = \frac{\mu_k - \mu(t_{k-1})}{h_\mu}$—the required average velocity vector determined at each step from the condition that the system reaches the state μ_k.

In fact, here we also solve the boundary value problem as in the case of the trajectory circuit described earlier. But, unlike the trajectory circuit, μ_k in principle, is not a free coordinate and is set based on the need to solve the problems before it by the trajectory circuit. The intensity of the control h_μ is set based on obtaining an acceptable quality of the transition process. In this case, since we are considering an isolated short-period motion, we will assume that μ_k is a free coordinate, and we solve the boundary value problem in the pilot circuit.

The next level of the hierarchy is represented by the circuit of angular velocities which is described by a system of equations

$$\dot{\omega} = \Omega(\omega\ ,\ M), \tag{6.174}$$

where $\omega = \left[\omega_x,\ \omega_y,\ \omega_z\right]^{\mathrm{T}}$ —state vector of the system;
$M = \left[M_x,\ M_y,\ M_z\right]^{\mathrm{T}}$—vector of the controls.
After the Taylor series expansion, the system (6.174) will take the form

$$\dot{\omega}_k = \Omega(\omega_{k-1}\ ,\ M_{k-1}) + \left.\frac{\partial\ \Omega}{\partial\ \omega^{\mathrm{T}}}\right|_{k-1} \Delta\ \omega_k + \left.\frac{\partial\ \Omega}{\partial\ M^{\mathrm{T}}}\right|_{k-1} \Delta\ M_k. \tag{6.175}$$

Hence, the increment of the vector of control at each step is determined by the following expression:

$$\Delta\ M_k = \left.\frac{\partial\ \Omega}{\partial\ M^{\mathrm{T}}}\right|_{k-1}^{-1} \left(\dot{\omega}_k - \Omega(\omega_{k-1}\ ,\ M_{k-1}) - \left.\frac{\partial\ \Omega}{\partial\ \omega^{\mathrm{T}}}\right|_{k-1} \Delta\ \omega_k\right), \tag{6.176}$$

where $\dot{\omega}_k = \left[\frac{\Delta\omega_{xk}}{h_\varpi}, \frac{\Delta\omega_{yk}}{h_\varpi}, \frac{\Delta\omega_{zk}}{h_\varpi}\right]^T$—the required average velocity vector, determined at each step from the need to obtain a solution for the Eq. (6.180);

h_ω—the intensity of control is determined by the additional condition.

The intensity of control determines the nature of the transition process and the level of its oscillation. Thus, changing the h_ω, we are able to get almost the entire range of transition process for a given object. In practice, the intensity of control is chosen either from the condition of obtaining an acceptable transition process, or from the condition of solving the optimization problem. It should be noted that, in principle, the intensity of control for each coordinate $\dot{\omega}_k$ can be selected individually depending on the flight mode, since the dynamics of the coordinates of the state vector of the Eq. (6.170) differ from each other. The resulting moment increments in Eq. (6.176) are converted to rudder deflections.

Consider the dynamic properties of the pilot circuit of the longitudinal channel. From the formula (6.172), it can be seen that the main parameter that affects the course of the transition process is the intensity of control h_μ. Exactly, this parameter determines the speed of reaching the given angle of attack in the longitudinal channel.

In Figs. 6.13, 6.14, 6.15, 6.16, 6.17 and 6.18, it is shown that the type of the transition process of the angle of attack and the nature of the deviation of the elevator for different values h_μ. The flight mode was selected as follows: H=1000 m, M=0.6. The task was to move the AC from the angle of attack equal to the balancing angle to the angle of attack equal to 8 deg. In Figs. 6.13 and 6.14, the problem was solved when $h_\mu = 1$ *s*., in Figs. 6.15 and 6.16—when $h_\mu = 0{,}7$ *s*, in Figs. 6.17 and 6.18—when $h_\mu = 0{,}3$ *s*. Hence, we can make a conclusion that by changing the intensity of the control, you can get almost any character of the transition process that is generally available for this object of the control.

The operation of the proposed algorithm of the control system, which is invariant to the failures of control and is based on the MDO principle, similar to the symmetry principle [40], was tested, using a numerical experiment by simulating the control processes of the AC with the standard ACS and with the proposed control system.

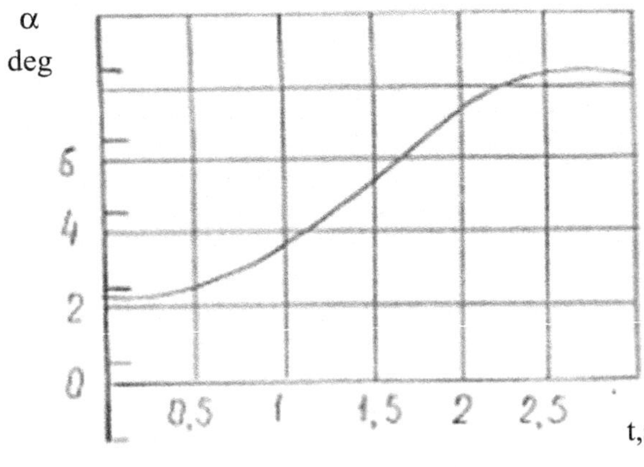

Fig. 6.13 The type of transition process of the angle of attack and the nature of the deviation of the elevator for the value $h_\mu = 1$ *s*

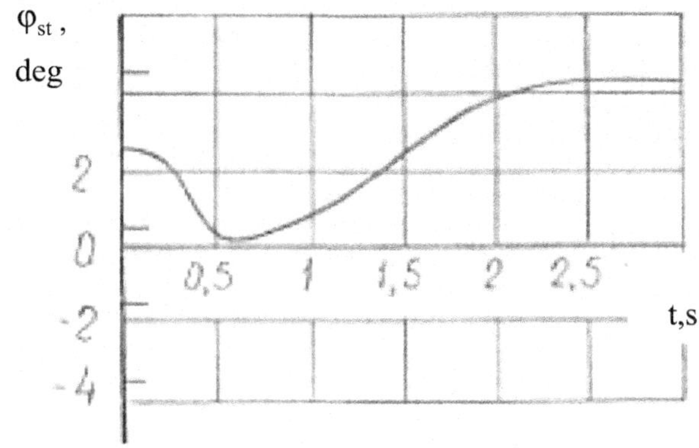

Fig. 6.14 The type of transition process of the angle of attack and the nature of the deviation of the elevator for the value h_μ = 1 *s*

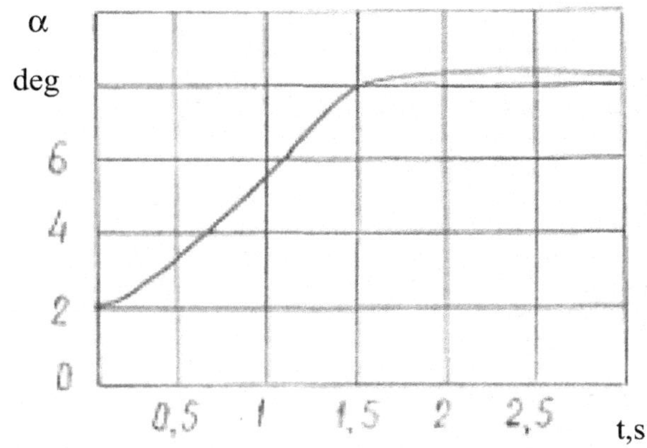

Fig. 6.15 The type of transition process of the angle of attack and the nature of the deviation of the elevator for the value h_μ= 0,7 *s*

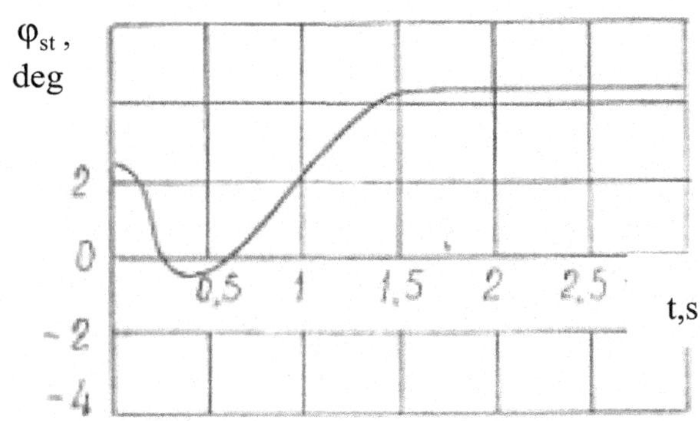

Fig. 6.16 The type of transition process of the angle of attack and the nature of the deviation of the elevator for the value h_μ = 0,7 *s*

The most critical failures of one of the console ofstabilizers from the point of view of flight safety were investigated as a failure. Its locks were considered in the balancing position with pitch control and in the position of $\varphi_r^{\text{fail}} = -5°$ (type failure «Withdrawal»).

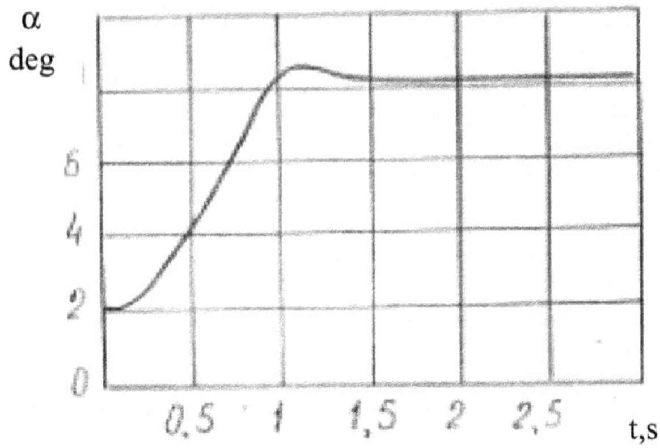

Fig. 6.17 The type of transition process of the angle of attack and the nature of the deviation of the elevator for the value $h_{\mu}=$ 0,3 *s*

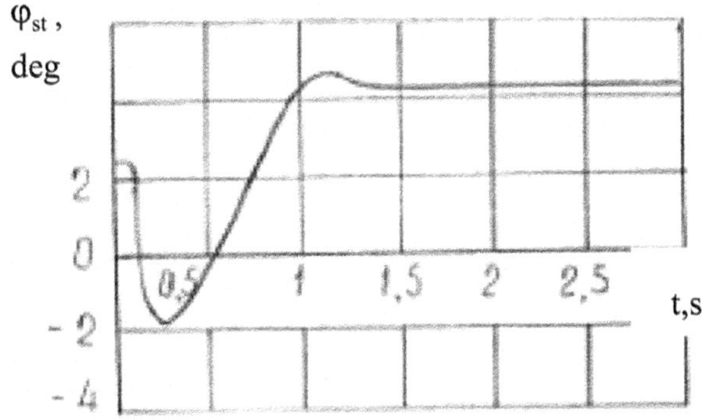

Fig. 6.18 The type of transition process of the angle of attack and the nature of the deviation of the elevator for the value $h_{\mu}=$ 0,3 *s*

It was assumed that the identification of the object parameters is performed simultaneously with the occurrence of a failure and with sufficient accuracy.

When the right half of the stabilizer is locked in the balancing position (Figs. 6.19, 6.20, 6.21, 6.22, 6.23, 6.24, 6.25, 6.26, 6.27, 6.28 and 6.29), the reaction of the AC with the standard ACS in the longitudinal channel is within acceptable limits. However, in the side channel, the roll and yaw speeds increase simultaneously. The application of the above algorithms eliminated these phenomena. The reaction of the AC was acceptable in all coordinates, the roll rate is close to zero. At the same time, the signal to the serviceable stabilizer increased, and the elements not involved in the standard ACS worked against the roll.

When the right stabilizer is moved to the position $\varphi_r^{\text{fail}} = -5°$ (Figs. 6.30, 6.31, 6.32, 6.33, 6.34, 6.35, 6.36, 6.37, 6.38, 6.39, 6.40 and 6.41), there is an intense increase in the angle of attack and overload with a significant drop in the angular velocity ω_x. At the same time, a significant static error in the motion parameters was observed. The elimination of this error requires appropriate actions of the pilot, which do not contradict his control stereotype.

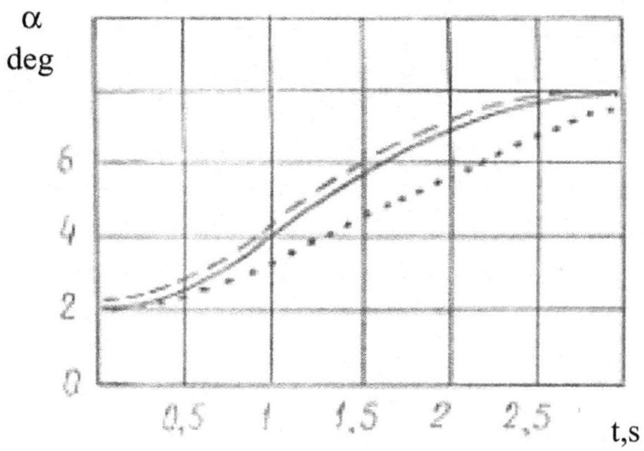

Fig. 6.19 The reaction of the AC with the standard ACS in the longitudinal channel when the right half of the stabilizer is blocked in the balancing position

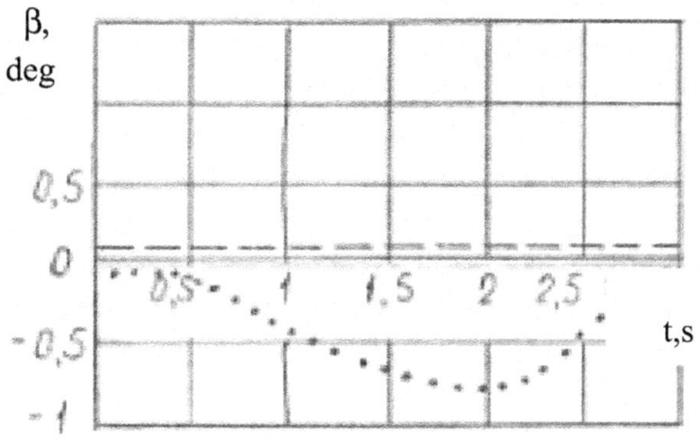

Fig. 6.20 The reaction of the AC with the standard ACS in the longitudinal channel when the right half of the stabilizer is blocked in the balancing position

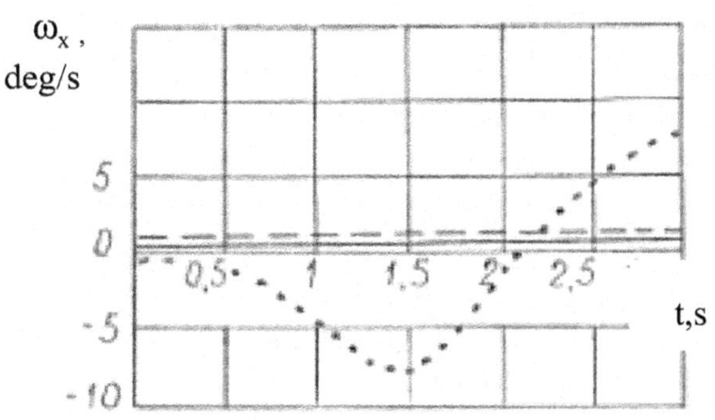

Fig. 6.21 The reaction of the AC with the standard ACS in the longitudinal channel when the right half of the stabilizer is blocked in the balancing position

It should be noted that in the previous sections, approximately identical results were obtained. Based on the analysis of the results, it can be concluded that this algorithm is not inferior in quality of the obtained transition processes to the algorithm with the reconfiguration of control systems. At the same time, we note that, in this

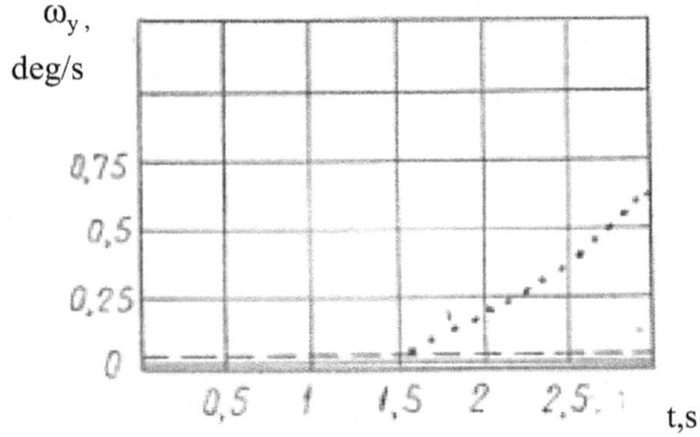

Fig. 6.22 The reaction of the AC with the standard ACS in the longitudinal channel when the right half of the stabilizer is blocked in the balancing position

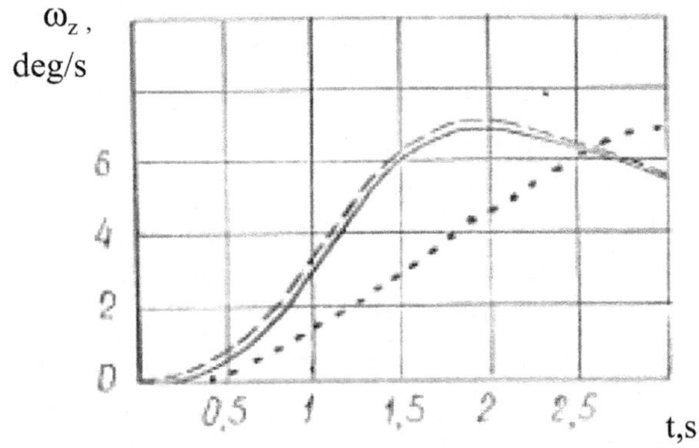

Fig. 6.23 The reaction of the AC with the standard ACS in the longitudinal channel when the right half of the stabilizer is blocked in the balancing position

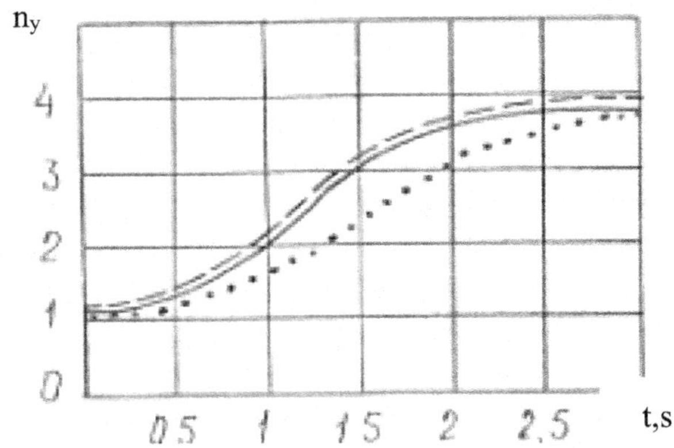

Fig. 6.24 The reaction of the AC with the standard ACS in the longitudinal channel when the right half of the stabilizer is blocked in the balancing position

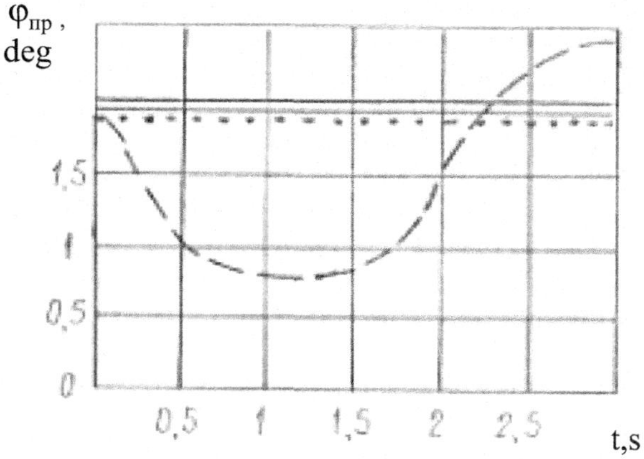

Fig. 6.25 The reaction of the AC with the standard ACS in the longitudinal channel when the right half of the stabilizer is blocked in the balancing position

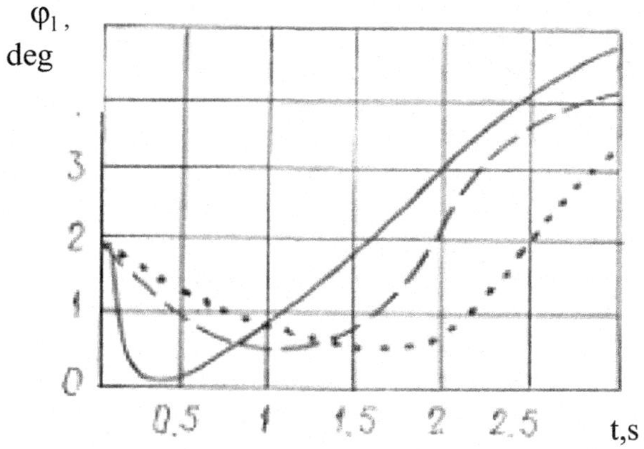

Fig. 6.26 The reaction of the AC with the standard ACS in the longitudinal channel when the right half of the stabilizer is blocked in the balancing position

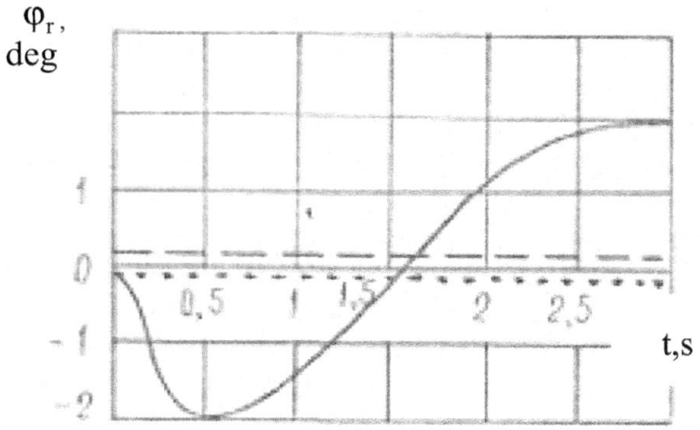

Fig. 6.27 The reaction of the AC with the standard ACS in the longitudinal channel when the right half of the stabilizer is blocked in the balancing position

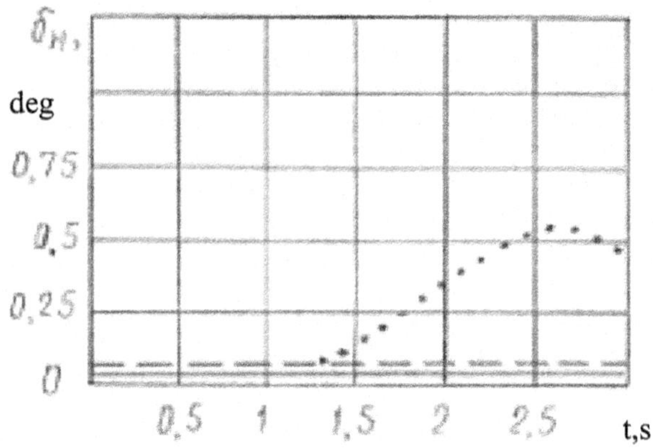

Fig. 6.28 The reaction of the AC with the standard ACS in the longitudinal channel when the right half of the stabilizer is blocked in the balancing position

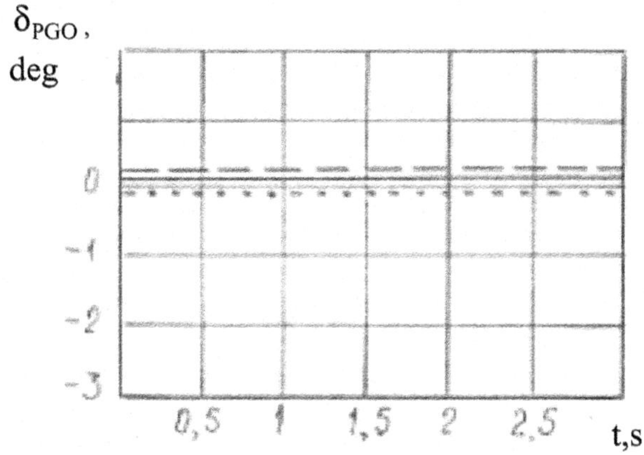

Fig. 6.29 The reaction of the AC with the standard ACS in the longitudinal channel when the right half of the stabilizer is blocked in the balancing position

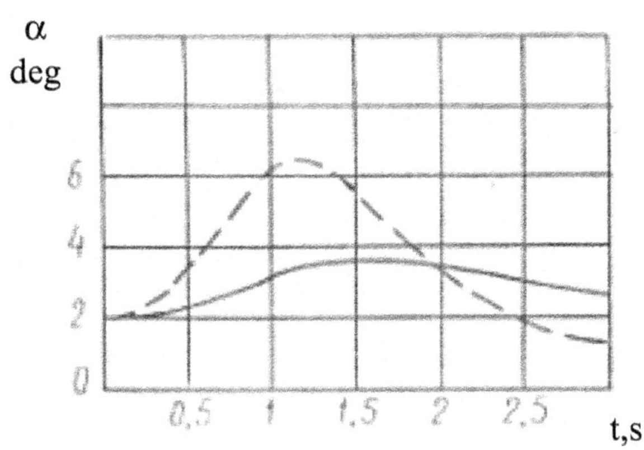

Fig. 6.30 Withdrawal of the right stabilizer to the position $\varphi_r^{fail} = -5°$

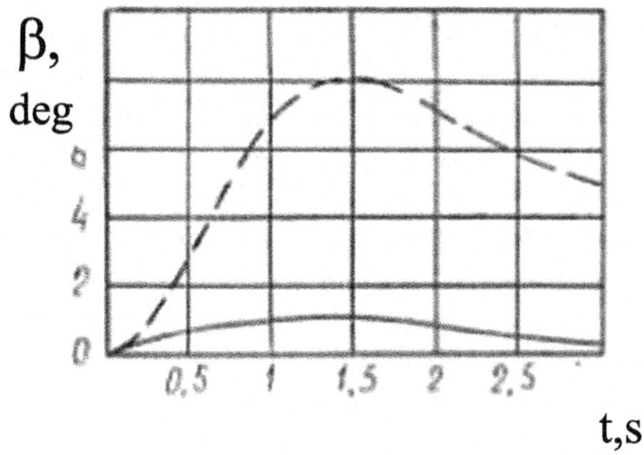

Fig. 6.31 Withdrawal of the right stabilizer to the position $\varphi_r^{\text{fail}} = -5°$

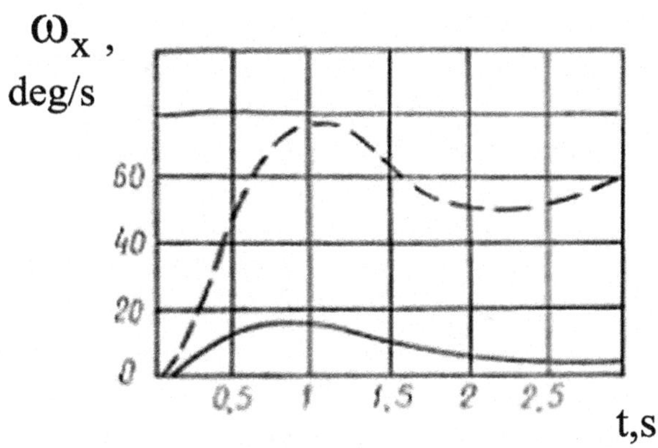

Fig. 6.32 Withdrawal of the right stabilizer to the position $\varphi_r^{\text{fail}} = -5°$

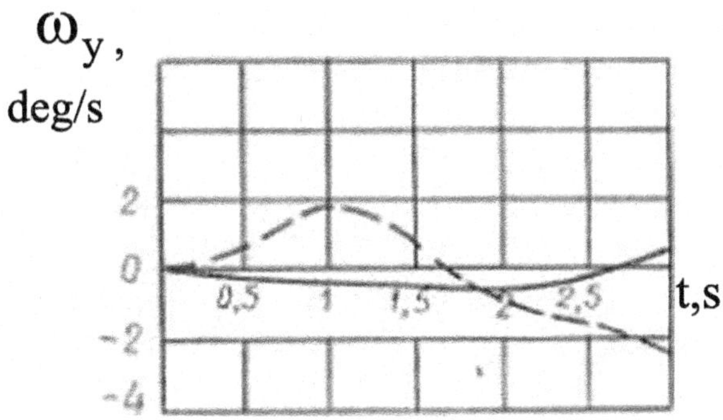

Fig. 6.33 Withdrawal of the right stabilizer to the position $\varphi_r^{\text{fail}} = -5°$

case, the control is based on a nonlinear model, taking into account all internal cross-links and the own characteristics of the AC.

The algorithm of reconfiguration does not provide for this, and the control is based on a linear AC model. In addition, these algorithms, in contrast to algorithms based on

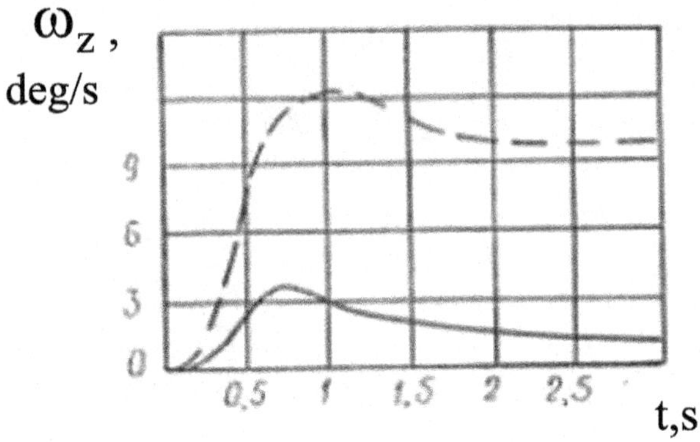

Fig. 6.34 Withdrawal of the right stabilizer to the position $\varphi_r^{\text{fail}} = -5^\circ$

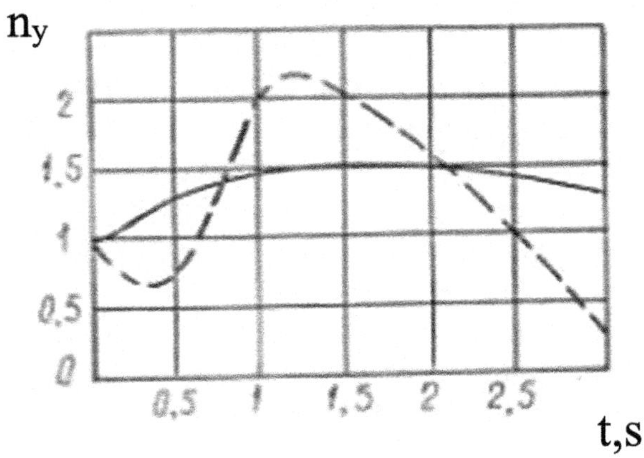

Fig. 6.35 Withdrawal of the right stabilizer to the position $\varphi_r^{\text{fail}} = -5^\circ$

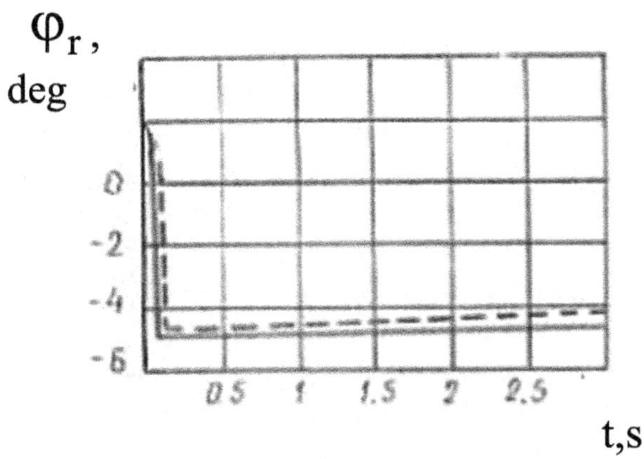

Fig. 6.36 Withdrawal of the right stabilizer to the position $\varphi_r^{\text{fail}} = -5^\circ$

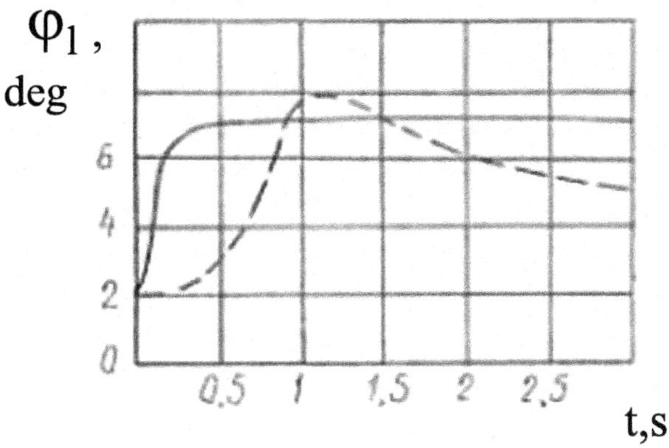

Fig. 6.37 Withdrawal of the right stabilizer to the position $\varphi_r^{\text{fail}} = -5°$

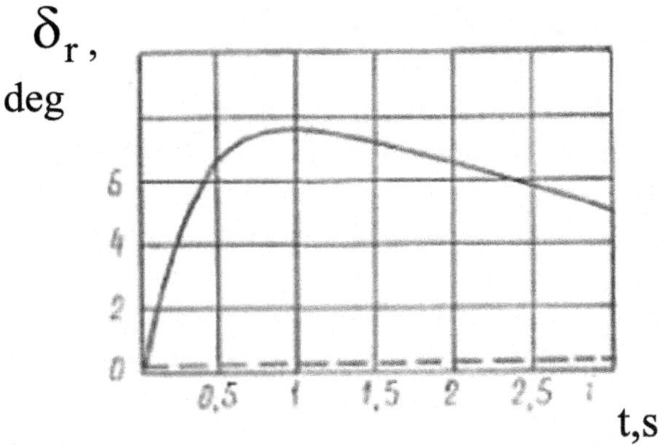

Fig. 6.38 Withdrawal of the right stabilizer to the position $\varphi_r^{\text{fail}} = -5°$

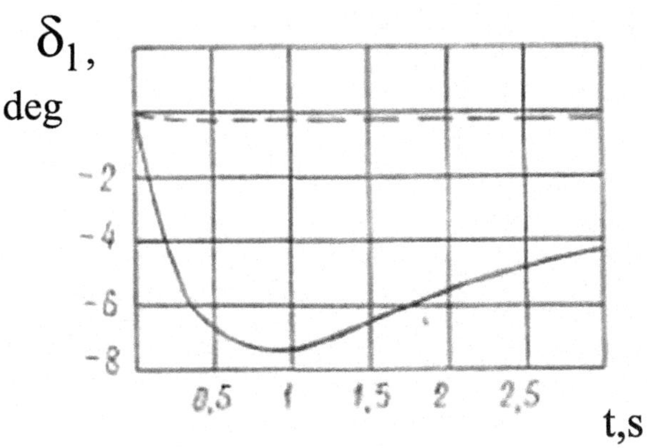

Fig. 6.39 Withdrawal of the right stabilizer to the position $\varphi_r^{\text{fail}} = -5°$

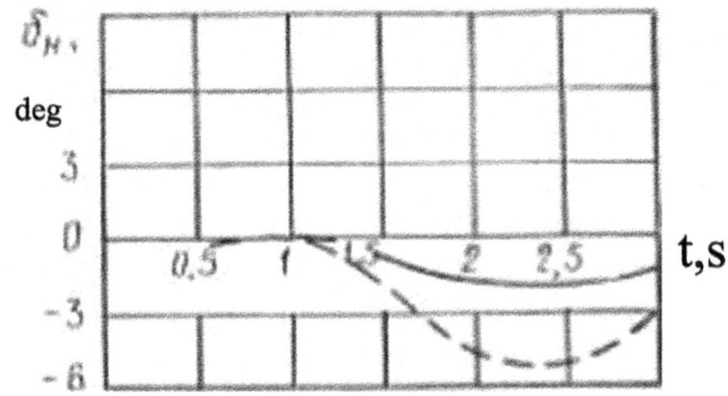

Fig. 6.40 Withdrawal of the right stabilizer to the position $\varphi_r^{\text{fail}} = -5°$

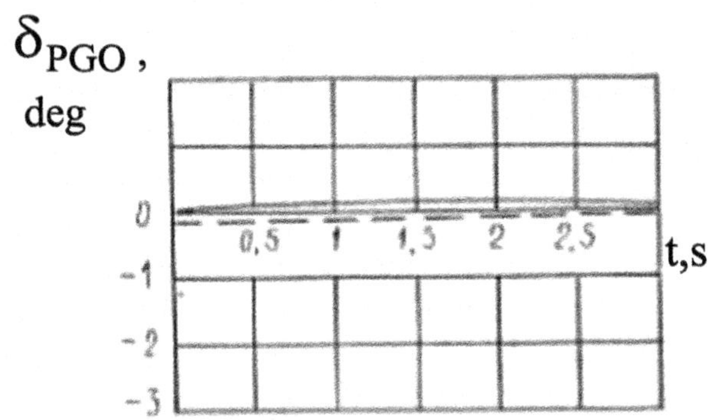

Fig. 6.41 Withdrawal of the right stabilizer to the position $\varphi_r^{\text{fail}} = -5°$

adaptive optimizing functional, are completely free from the effect of «readaptation», and require about an order of magnitude less required performance of the on-board digital computing system (OBDCS).

6.6 Technique and Features of Aircraft Thrust Vector Control

The ability to control the thrust vector of the AC opens up new prospects in the field of expanding the dynamic properties of the AC, as well as improving the level of flight safety. At the same time, the implementation of this feature is an extraordinary task in the sense of solving it with the help of a standard control system of the AC. The reason is that the control of the movement of the AC through a change in the direction of the thrust vector has its effect simultaneously on several coordinates of the state vector of the AC and with a strongly pronounced nonlinear character. In addition, the cross-links of these coordinates, in particular the velocity and angle of attack, are also represented by nonlinear dependencies.

The book offers a solution to this problem by using an algorithm of control that implements the ideas of the MDO. As an example, we will consider the solution of the problem only for the longitudinal channel, but it seems no less effective to use a controlled thrust vector in the lateral channel.

When describing the full model of the AC, we will limit ourselves only to the equations describing the short-period movement of the AC in the longitudinal channel with the addition of the velocity equation. Then, taking this into account, the system of equations, describing the dynamics of the controlled thrust vector, are represented as

$$
\begin{aligned}
\dot{\alpha} &= \varpi_z - \frac{g}{V}\left[\left(n_x \sin\, \alpha + n_y \cos\, \alpha\right) - (\sin\, \vartheta\, \sin\, \alpha + \cos\, \vartheta\, \cos\, \alpha)\right]; \\
\dot{V} &= g\left\{n_x \cos\alpha - n_y \sin\alpha - \sin\vartheta \cos\alpha + \cos\vartheta \sin\alpha\right\}; \\
\dot{\omega}_2 &= \frac{1}{I_{zz}}\left[M_z - R\, y_r\, \sin\, \varphi_r\right]; \quad \dot{\upsilon} = \omega_z \\
n_x &= \frac{1}{m\, g}(R\, \cos\, \varphi_r - X)\, ; \\
n_y &= \frac{1}{m\, g}(R\, \sin\, \varphi_r + Y)\, ;
\end{aligned}
\tag{6.177}
$$

where y_r—the distance to the point of application of the resultant thrust gain, when the thrust vector deviates from the neutral position.

The system of Eqs. (6.177) presents the equations and relationships, that are most strongly affected by the deviation of the AC thrust vector.

In order to use the MDO algorithms, it is necessary to find out which parameters on the right sides of the Eqs. (6.177) are directly controlling the coordinates α and v, and which ones have a side effect.

The analysis shows that for the angle of attack α, the control coordinate is the angular velocity ϖ_z, all other components of the right side in the equation on the angle of attack determine either the proper movement of the AC or the cross-links with other channels. A similar conclusion can be drawn for the equation of velocity. Here, the main control effect is overload n_x.

If we limit ourselves to considering only the control functions of the deflected thrust vector of the AC, then from the equations for angular velocity φ_z and overload n_x, the system (6.177) can be seen that these coordinates can be controlled by changing either the thrust force R or the angle of deviation of the thrust vector φ_r. At the same time, the effect of these control parameters on the angular velocity and overload is strongly nonlinear. In addition, the relationship of the equations for the angle of attack and the speed of the system (6.177) between themselves by n_x, n_y, α, is also expressed by nonlinear dependencies.

To compensate such «dual» nonlinear effect allows the algorithm of control, using the MDO principle.

6.6.1 Aircraft Thrust Vector Control Algorithm

According to the principles of the building of controls outlined above, we present the first two equations of the system (6.177) by the following system of equations:

$$\dot{z} = f(z\ ,\ u), \tag{6.178}$$

where $z = [\alpha,\ V]^{\mathrm{T}}$—state vector of the system;

$u = \left[\varpi_z,\ n_x\right]^{\mathrm{T}}$—vector of control.

After the decomposition of the system (6.178) in Taylor's row, and using only the first members of the row, we get

$$\dot{z}_{\mathrm{fin}} = f(z_{\mathrm{fin}-1}) + \left.\frac{\partial\ f}{\partial\ z^{\mathrm{T}}}\right|_{fin-1} \Delta\ z_{\mathrm{fin}} + \left.\frac{\partial\ f}{\partial\ u^{\mathrm{T}}}\right|_{\mathrm{fin}-1} \Delta\ u_k \tag{6.179}$$

The second term of the right side of the Eq. (6.179) shows the proper movement of the AC, the third term shows the forced movement. Then, from Eq. (6.179), we can obtain the value of the vector of increments of control actions as a function of the vector of the state of the system (6.179) and its first derivative in the case, if such exists. Then the vector of increment of controls at each step will be determined by the expression

$$\Delta\ u_{\mathrm{fin}} = \left.\frac{\partial\ f}{\partial\ u^{\mathrm{T}}}\right|_{fin-1}^{-1} \left(\dot{z}_{\mathrm{fin}} - \left.\frac{\partial\ f}{\partial\ z^{\mathrm{T}}}\right|_{\mathrm{fin}-1} \Delta\ z_{\mathrm{fin}}\ -\ f(y_{\mathrm{fin}-1})\right), \tag{6.180}$$

where $\frac{\partial f}{\partial z^{\mathrm{T}}} = \begin{bmatrix} \frac{\partial f_1}{\partial \alpha} & \frac{\partial f_1}{\partial V} \\ \frac{\partial f_2}{\partial \alpha} & \frac{\partial f_2}{\partial V} \end{bmatrix}$—matrix of own properties of the object;

$\frac{\partial f}{\partial u^{\mathrm{T}}} = \begin{bmatrix} \frac{\partial f_1}{\partial \omega_z} & \frac{\partial f_1}{\partial n_x} \\ \frac{\partial f_2}{\partial \omega_z} & \frac{\partial f_2}{\partial n_x} \end{bmatrix}$—matrix of the effectiveness of controls.

The derivative of the state vector in the expression (6.180) can be represented as $\dot{z}_{\mathrm{fin}} = \frac{z_{\mathrm{fin}} - z(t_{\mathrm{fin}-1})}{T - t_{\mathrm{fin}-1}}$—the required average velocity vector, determined at each step from the condition of the system reduction (6.170) to the point z_k for the time $T - t_k$ from the current state of the system.

Usually, for a short period of movement, a constant speed value is required, and the final value of the angle of attack is set, based on the conditions of the given task. The final time T is determined from the given characteristics of the transition process.

The received values of the increment of overload n_x are converted to the necessary increment of the thrust of the power device of the AC, adjusted for deviations of the thrust vector from the construction axis of the AC. Then the thrust is determined at each step by the following recurrent expression

$$\Delta\ R_{\text{fin}} = \frac{\Delta\ n_{x_{\text{fin}}}\ m\ g - X}{\cos\ \phi_{r_{\text{fin}}}}; \tag{6.181}$$

$$R_{\text{fin}+1} = R_{\text{fin}} + \Delta R_{\text{fin}} \tag{6.182}$$

Now let's take a closer look at how you can get an increment in the angular velocity φ_z

$$\dot{\omega}_z = \frac{1}{I_{zz}}\left[M_z - R\ y_r\ \sin\ \varphi_r\right]. \tag{6.183}$$

The analysis of Eq. (6.183) follows that the forced movement of the angular velocity φ_z is determined by three components. The first component is the moment perturbation φ_r, which is determined by the flight mode, the features of the aerodynamic configuration and the deviation of the stabilizer and the front horizontal tail; the second component is the thrust of the engine *R*; the third component is the angle of deviation of the engine thrust vector from the construction axis of the AC φ_r. Here it should be noted that the engine thrust and angle φ_r are included in the Eq. (6.183) in a nonlinear way, which is the reason for the difficulties of using the previously considered algorithms, and, even more so, the standard control system for controlling the angle of attack by the deviations of the thrust vector. In addition, the engine thrust is a control parameter only, if φ_r. Let us present Eq. (6.183) as

$$\dot{\omega}_z = \xi(M_z\ ,\ R\ ,\ \varphi_r). \tag{6.184}$$

We decompose Eq. (6.184) into Taylor's row using only the higher terms of row

$$\dot{\omega}_{z_{\text{fin}}} = \xi_{\text{fin}-1} + \left.\frac{\partial\ \xi}{\partial\ M_z}\right|_{fin-1} \Delta\ M_{z_k} + \left.\frac{\partial\ \xi}{\partial\ R}\right|_{\text{fin}-1} \Delta\ R_{\text{fin}} + \left.\frac{\partial\ \xi}{\partial\ \phi_r}\right|_{\text{fin}-1} \Delta\ \phi_{r_k}. \tag{6.185}$$

Let's present Eq. (6.185) in matrix form

$$\dot{\omega}_{z_{fin}} = \xi_{fin-1} + \left[\ \left.\frac{\partial\ \xi}{\partial\ M_z}\right|_{fin-1}\ \ \left.\frac{\partial\ \xi}{\partial\ R}\right|_{fin-1}\ \ \left.\frac{\partial\ \xi}{\partial\ \phi_r}\right|_{fin-1}\ \right]^T \begin{bmatrix} \Delta\ M_z \\ \Delta\ R \\ \Delta\ \phi_r \end{bmatrix}, \tag{6.186}$$

Let's introduce the notation $\Delta\tilde{u} = \left[\Delta M_z\ \ \Delta R\ \ \Delta\varphi_r\right]^T$—increment of the control vector of the Eq. (6.186);

$\left.\frac{\partial\xi}{\partial\tilde{u}^T}\right|_{\text{fin}-1} = \left[\ \left.\frac{\partial\xi}{\partial M_z}\right|_{\text{fin}-1}\ \ \left.\frac{\partial\xi}{\partial R}\right|_{\text{fin}-1}\ \ \left.\frac{\partial\xi}{\partial\varphi_r}\right|_{\text{fin}-1}\ \right]^{\text{T}}$—vector of the control's efficiency.

Then the increment of the vector of controls will be determined by the formula

$$\begin{bmatrix} \Delta M_z \\ \Delta R \\ \Delta \phi_r \end{bmatrix} = \left[\left.\frac{\partial \xi}{\partial M_z}\right|_{\text{fin}-1} \left.\frac{\partial \xi}{\partial R}\right|_{fin-1} \left.\frac{\partial \xi}{\partial \phi_r}\right|_{\text{fin}-1} \right]^{\#} \left(\dot{\omega}_{z\text{fin}} - \xi_{\text{fin}-1}\right), \tag{6.187}$$

where $\dot{\varpi}_{z\,\text{fin}} = \frac{\Delta u_1}{h_\omega}$—the required average velocity vectoris determined at each step from the need to obtain a solution of the equation by the angle of attack.

The increment of aerodynamic torque $\Delta\ M_z$ we get from the equation, that determines the moment M_z

$$M_z = M_z(\ \phi_{st}\ ,\ \delta_{\text{pgo}}\ ). \tag{6.188}$$

If we assume that the deflection of the rudders occurs instantly, then it follows from the linear nature of the Eq. (6.187) that the increment of moments is also step by step type. In reality, this circuit has the character of an inertial link, but since its time constant is much smaller than that of the other levels of the hierarchy, this assumption seems plausible. This will a bit simplify the research and focus attention on the dynamics of the AC omitting the dynamics of the executive devices of the control system. Although with the known characteristics of the transfer functions of the executive devices, it is easy to take them into account by changing the nature of the system of Eq. (6.177) and applying the proposed method.

The system of Eq. (6.187) can be written so that the increment of the rudder deflection vector enters the right side of the system in a linear way

$$\Delta\ M_z = \left.\frac{\partial\ M_z}{\partial\ \phi_{st}^T}\right|_{\text{fin}} \Delta\ \phi_{\text{st}\cdot\text{fin}} + \left.\frac{\partial\ M_z}{\partial\ \delta_{\text{pgo}}^{\text{T}}}\right|_{\text{fin}} \Delta\ \delta_{\text{pgo}_{\text{fin}}}, \tag{6.189}$$

where $\frac{\partial M_z}{\partial \phi_{st}}$—efficiency of the stabilizer by torque M_z;

$\frac{\partial M_z}{\partial \phi_{st}}$—efficiency of the front horizontal tail by torque M_z.

Then the increment of the rudder deflection vector is obtained by the inverse transformation from the system of Eqs. (6.188)

$$\cdot\begin{bmatrix} \Delta\phi_{st_{fin}} \\ \Delta\delta_{PGO_{fin}} \end{bmatrix} = \left[\left.\frac{\partial M_z}{\partial \phi_{st}}\right|_{fin} \left.\frac{\partial M_z}{\partial \delta_{PGO}}\right|_{fin} \right]^{\#} \cdot \Delta M_z. \tag{6.190}$$

So, using the MDO, we obtained from the system of Eqs. (6.188) the expressions for four controls: the engine thrust, the stabilizer, the front horizontal tail, and the angle of deviation of the engine thrust vector.

Now let's analyze the expressions (6.184)–(6.190) for a clear understanding of the physics of the interaction of these controls and the two coordinates to be controlled: the angle of attack and the speed of the AC.

To simplify the analysis, we will first consider in more detail the available methods of controlling the AC and their possible combinations. From Eqs. (6.178)–(6.191), it follows that the circuit of the angle of attack is a double integrating link, and the angle of attack itself is not directly affected by the vector of control (6.187) (except for weak relationships, which can be ignored to simplify the analysis). In fact, all the forced movements in the differential equation, describing the angle of attack, are determined by the angular velocity ω_z. Equation (6.186) just describes the influence of the control devices ω_z. Let's look at it in more detail.

From Eqs. (6.182)–(6.190), it follows that the angular velocity can be controlled by several devices of controls at once, namely the stabilizer, the front horizontal tail, the engine thrust together with the engine thrust vector deflection system, and purely the engine thrust vector deflection angle. Here we have a case of an excess of controls. On the one hand, this makes it possible to significantly expand the dynamic capabilities of the system and increase the level of flight safety by duplicating the devices of controls, on the other hand, the task is to optimally use the existing excess of the control action area. In this case, the problem is that the available devices of controls have an effect on the angular velocity ω_z, which sometimes differs by several orders of magnitude. In addition, the efficiency of the impact of the thrust and the angle of deviation of the thrust vector have a strongly pronounced nonlinear character determined by the trigonometric dependence on the angle of deviation of the thrust vector.

The laws of control (6.187), (6.200) do not take into account the physical restrictions on the control, which leads to unjustified losses of efficiency, since some devices can go to the restrictions, while others, having less efficiency at a given coordinate, will be far away from them. In order to take into account the physical constraints on the controls and the differences in the effectiveness of the devices of control, the norm of additional deviation of the control vector $\Delta\,\tilde{u}$ is minimized by normalizing its physical limits

$$\Delta\,\tilde{u}_N = D^{-1}\Delta\,\tilde{u}, \qquad (6.191)$$

where $\Delta\,\tilde{u}_N$—the normalized vector of the control's increment;

D^{-1}—diagonal matrix of constraints on the deviation of control surfaces.

In this case, when the matrix pseudo-rotation operation is used, the load on the handlebars will be distributed according to their efficiency and the range of deviations.

6.6.2 *Analysis of the Performance of the Thrust Vector Control Algorithm*

It was previously shown that the law (6.180) and (6.187) allows us also to control the angle of attack by changing the value of the engine thrust. In theoretical terms, if we do not take into account the restrictions on the speed of change in the engine thrust, we get a fairly powerful component of the control of angle of attack, which could take on itself a significant part of the load in the case of control of this coordinate in condition, that the angle of deviation of the thrust vector from the construction axis of the AC is not zero. But, in the case of modeling the processes of control, taking into account all the real restrictions on the range and speed of deviation of the devices of control which are used, we get a significant fluctuation in the deviation of all devices of control and, as a result, in the change of the angle of attack.

The conducted analysis shows that the oscillation of the processes occurs due to a significant difference in the rates of change of the actual engine thrust, the angular positions of the stabilizer, and the direction of the engine thrust vector. These differences are determined by the physical nature of these devices and are not subject to any change for the control object under study. Although it should be noted that with a fairly slow change in the angle of attack, the thrust can still take on part of the load if this does not contradict the acceleration of the engines.

From this, we can conclude that it is hardly advisable to use the engine thrust increment to control the angle of attack. But the engine thrust must be used to control the speed of the AC. Note that the speed of the AC, as it is known, can be controlled by vertical overload. In this case, we will limit ourselves only to the engine thrust, since vertical overload is tied to the control of the angle of attack, and it is simply unacceptable to control various parameters such as speed and angle of attack, with the help of identical devices of control, in some modes.

Taking into account the above, we will consider the simulation of the transition process by the angle of attack from the balancing position to the given one, if it is necessary to preserve the speed of the AC. The flight mode was chosen as follows: $H = 1000$ m, $M = 0.6$. The task was to transfer the AC from the angle of attack equal to the balancing angle to the angle of attack equal to 8 deg, using only the angle of deviation of the thrust vector for control. The stabilizer remains in the balancing position.

In Figs. 6.42, 6.43, 6.44 and 6.45, the simulation results are presented. As can be seen from these figures, the angle of attack reaches the set value and stabilizes on it, while the speed of the aircraft remains constant, and the necessary increment of the longitudinal overload is worked out by the engine thrust. At the same time, there is no oscillation, and the transition process in the angle of attack corresponds to the existing requirements.

Using the operation of normalization of control actions, modeling of various versions of load distribution in the control of the AC by the angle of attack on various controls was carried out. The flight mode and control task are identical to the previous ones. Figures 6.46, 6.47, 6.48 and 6.49 show different load distribution options when performing the same task. Three devices of the control are used: the stabilizer, the front horizontal tail, and the controlled thrust vector of the AC. Artificially changing the relative efficiency of a particular device of control, we have the opportunity to influence the process of redistributing the load between them. This allows us to expand significantly the range of control actions and improve flight safety.

Now let's look at how the proposed algorithm works in case of failures of the control devices. When the right half of the stabilizer is locked in the balancing position during pitch control (Figs. 6.50 and 6.51), the roll and yaw speeds increase simultaneously in the side channel.

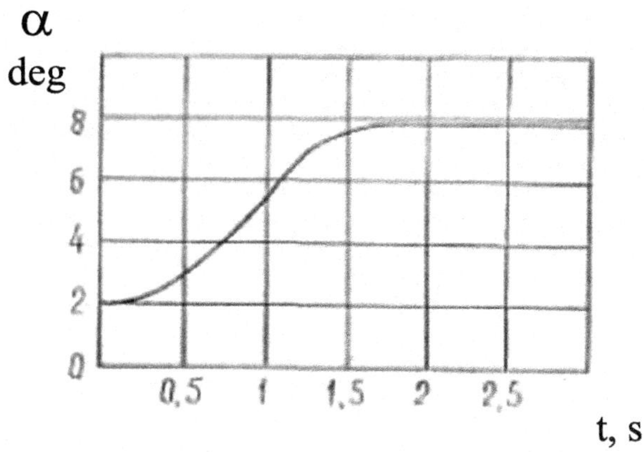

Fig. 6.42 The simulation of the transition process by the angle of attack from the balancing position to the given one if it is necessary to preserve the speed of the AC

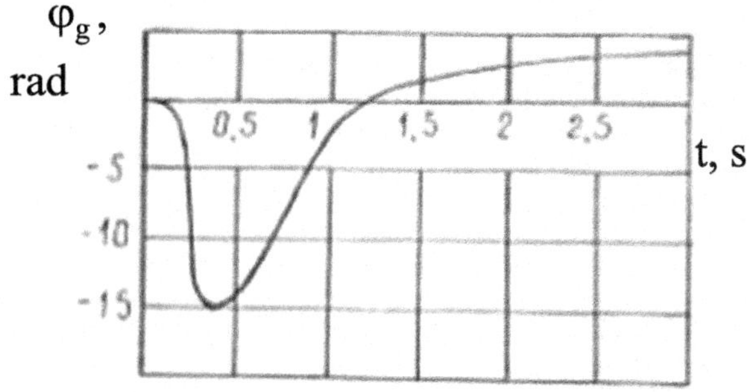

Fig. 6.43 The simulation of the transition process by the angle of attack from the balancing position to the given one if it is necessary to preserve the speed of the AC

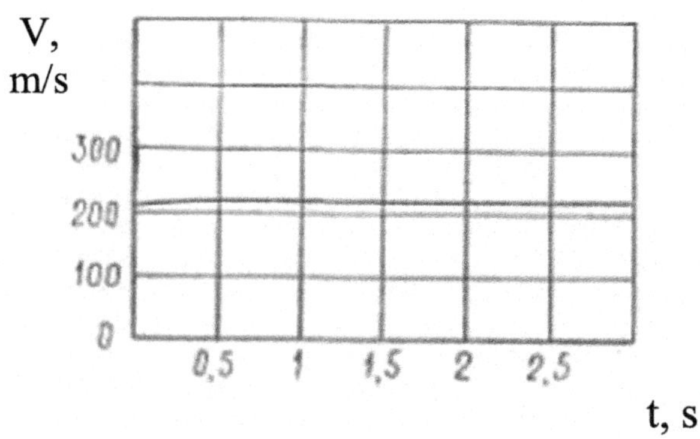

Fig. 6.44 The simulation of the transition process by the angle of attack from the balancing position to the given one if it is necessary to preserve the speed of the AC.

The use of the above algorithms eliminated these phenomena. The reaction of the AC was acceptable for all coordinates, the roll rate is close to zero. At the same time, the signal to the serviceable stabilizer increased, and part of the load in the longitudinal channel was taken over by the channel of the controlled thrust vector, the elements not involved in the regular ACS worked against the roll.

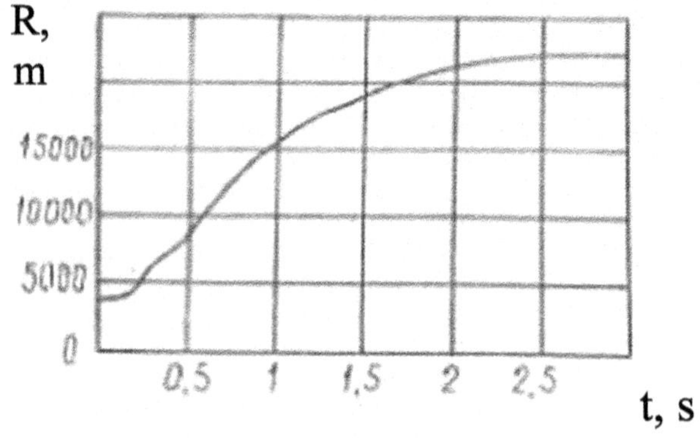

Fig. 6.45 The simulation of the transition process by the angle of attack from the balancing position to the given one if it is necessary to preserve the speed of the AC

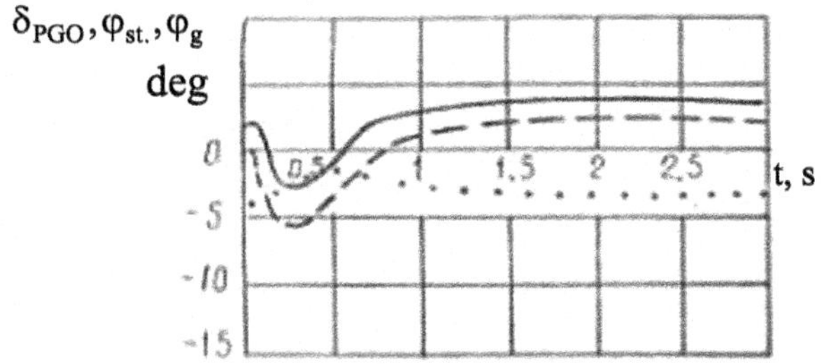

Fig. 6.46 An option of load distribution when performing the same task

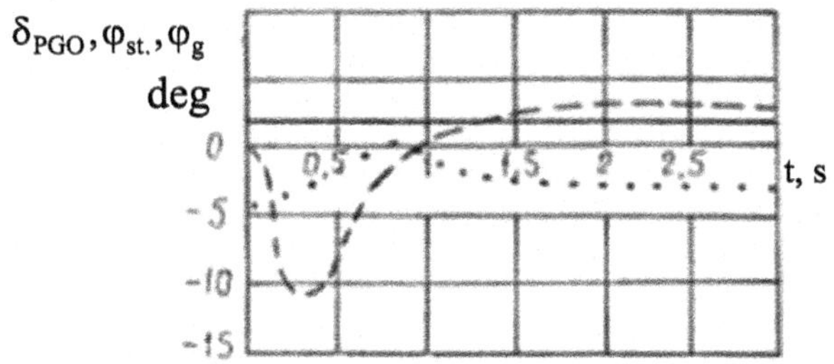

Fig. 6.47 An option of load distribution when performing the same task

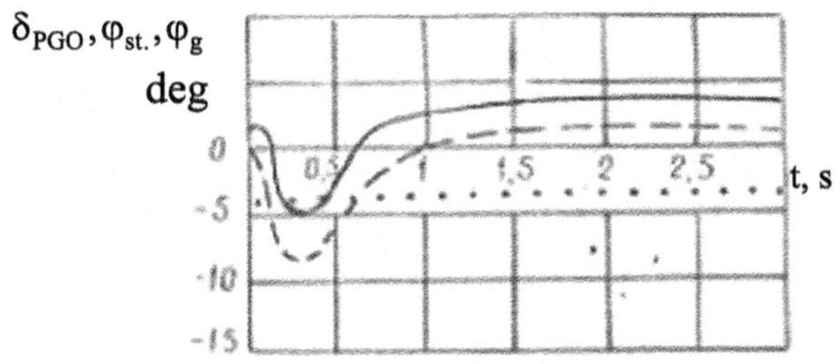

Fig. 6.48 An option of load distribution when performing the same task

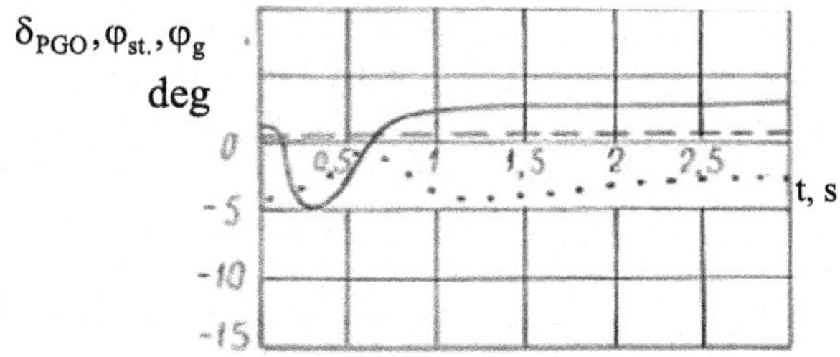

Fig. 6.49 An option of load distribution when performing the same task

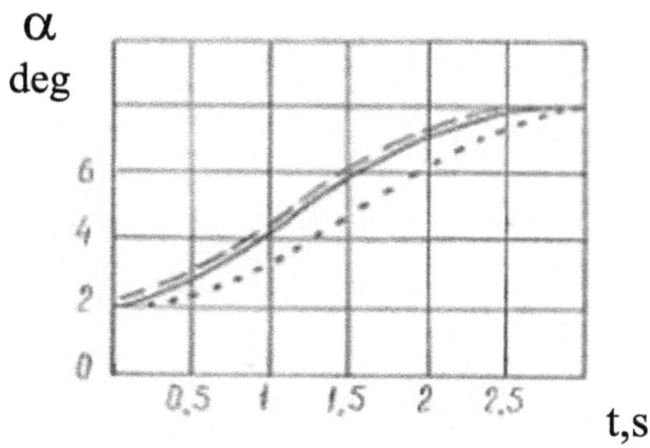

Fig. 6.50 The case of locking the right half of the stabilizer in the balancing position during pitch control

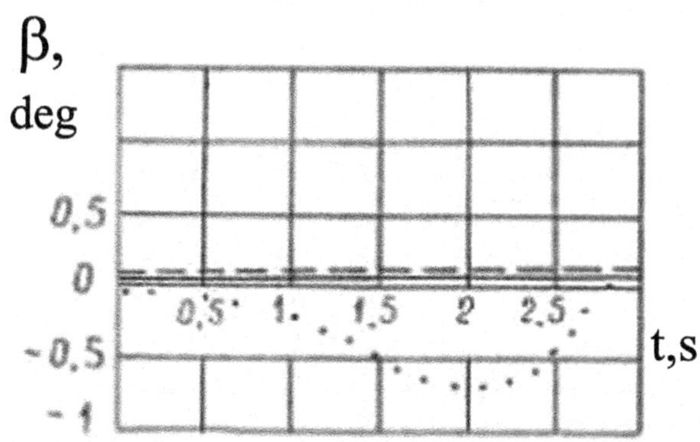

Fig. 6.51. The case of locking the right half of the stabilizer in the balancing position during pitch control

The use of a controlled thrust vector may have a specific interest in the case of flying at low speeds, especially during pre-landing maneuvers. The flight mode was selected as follows: $H = 1000$ m, $M = 0.2$. The task was to transfer the AC from the angle of attack equal to the balancing angle to the angle of attack equal to 10 deg, using the following combinations for control: only the stabilizer; only the angle of deviation of the thrust vector; their joint operation.

Figures 6.52, 6.53, 6.54, 6.55, 6.56 and 6.57 show the results of the simulation. The analysis of the results shows that at low speed, the stabilizer is not able to solve the task and stabilize the AC at a given angle of attack since its effectiveness largely depends on the speed rush, and at low flight, speed is quite low.

In the case of using only the deflected thrust vector of the AC, the problem of the angle of attack is solved, but with significant deviations of the thrust vector, the speed of the AC decreases, because the thrust machine cannot quickly react to the loss of the value of thrust projection on the velocity vector.

When the devices of control work together, the reaction of the AC, both in terms of speed and angle of attack, is within acceptable limits.

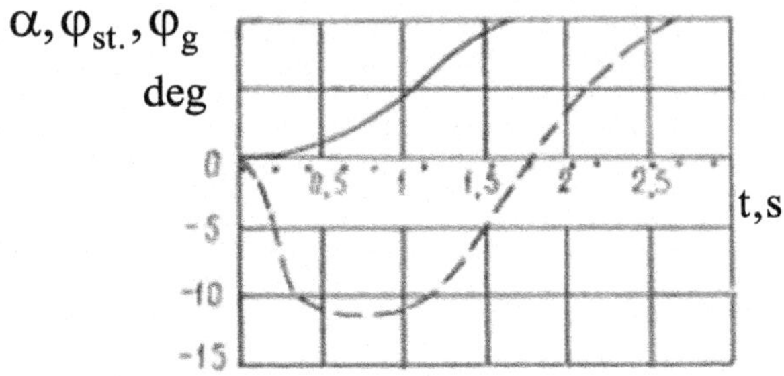

Fig. 6.52 Results of simulation of the use of the controlled thrust vector in low speed flights

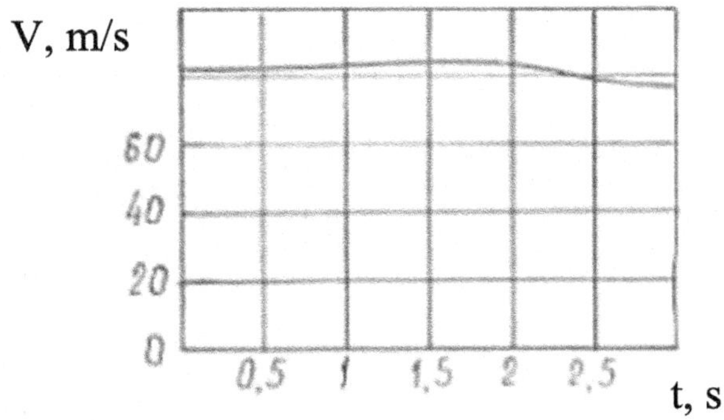

Fig. 6.53 Results of simulation of the use of the controlled thrust vector in low speed flights

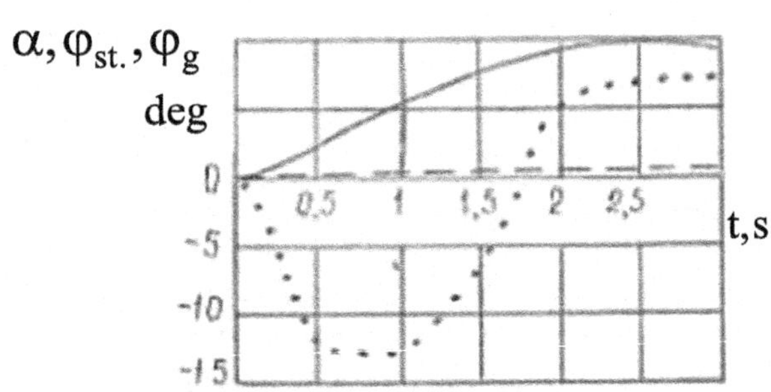

Fig. 6.54 Results of simulation of the use of the controlled thrust vector in low speed flights

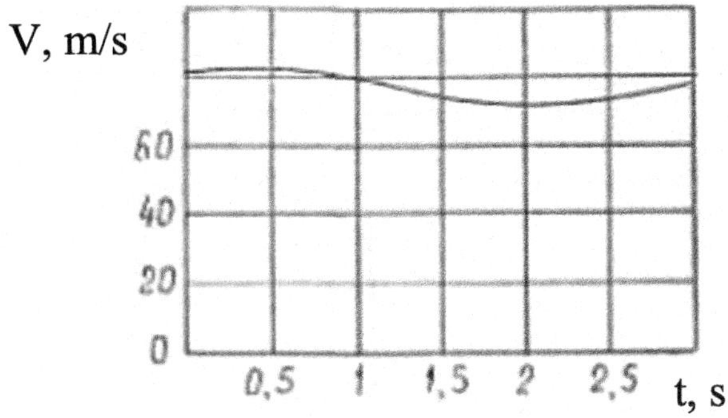

Fig. 6.55 Results of simulation of the use of the controlled thrust vector in low speed flights

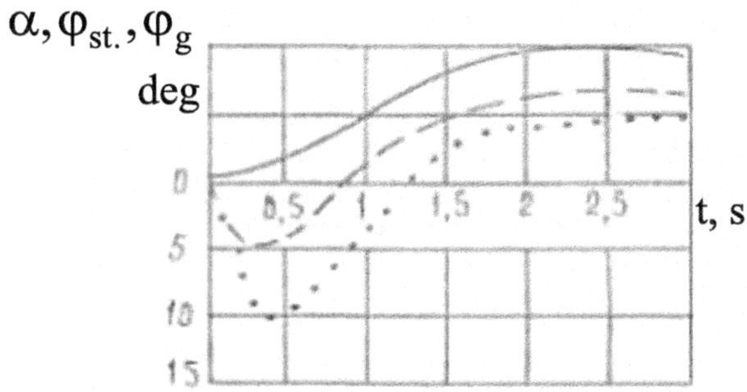

Fig. 6.56 Results of simulation of the use of the controlled thrust vector in low speed flights

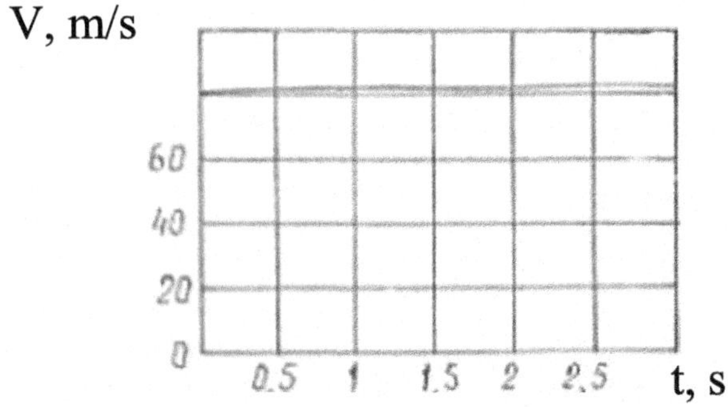

Fig. 6.57 Results of simulation of the use of the controlled thrust vector in low speed flights

6.7 Extended Version of the Method of Direct Optimization

In accordance with the accepted (item 6.3.3.2) scheme for generating the required values of the derivatives of the state vector and the nature of the applied problems to be solved, MDO forms the shortest trajectories of the AC movement to the given areas of space.

In many cases, as the results of testing the MDO on solving the simplest tasks of optimal control, such a scheme provides to receive a solutions, that completely coincide with the optimal ones, obtained by classical methods.

In the case of optimization of the spatial movement of the AC, the method also gives results, that coincide with the trajectory, obtained on the basis of the principle of maximum.

This coincidence can be explained by the formulation of such problems itself, mainly by optimization criteria, boundary conditions, and constraints on the state and control vectors of the object. It is even intuitively clear that on the shortest trajectories, you can get the minimum time, the minimum fuel consumption, etc. even within the framework of the accepted scheme of formation and variation of the required values of the derivatives of the state vector.

On the other hand, it is also clear that there are problems beyond the frames of this book, in which the optimal solution does not belong to the shortest trajectories. To solve such problems, you need to use the extended version of the MDO.

We formulate the principle on which the functioning of the algorithms of the extended version of the MDO is based.

«The trajectory is not optimal if there is at least one of its sections that can be improved by maintaining continuity at its ends along the coordinates of the state and control vectors with adjacent sections of this trajectory».

The validity of this statement is obvious. However, the scheme for improving the trajectory sections is far from a trivial task, especially for control objects, described by nonlinear differential equations.

We will present at the beginning an extended version of the MDO for the simplest object—the double integrating link

$$\begin{aligned} \dot{x} &= y, \quad x(t_0) = x_0, \quad y(t_0) = y_0, \quad u(t_0) = u_0; \\ \dot{y} &= u, \quad x(t_{\text{fin}}) = x_{\text{fin}}, \quad y(t_{\text{fin}}) = y_{\text{fin}}, \quad u(t_{\text{fin}}) = u_{\text{fin}}; \\ I &= \int_{t_0}^{t_{\text{fin}}} Q(x, u, t)dt. \end{aligned} \tag{6.192}$$

Let's assume that the basic trajectory of the movement of this object $\hat{x}(t), \ \hat{y}(t), \ \hat{u}(t)$ is obtained. We will divide this trajectory into arbitrary sections as shown in Fig. 6.58. Consider the technology for improving the trajectory $\hat{x}(t)$.

First, we select a section of the trajectory on the interval $[t_0, t_2]$. In the point t_1, let's implement a variation $\hat{x}(t)$ on value δx, then $x(t_1) = \hat{x}(t_1) + \delta x$.

To determine the derivatives at a point t_1, we will proceed as follows.

Let's define the equation of the trajectory section $\varphi(x) = 0$ in the form of the equation of a parabola passing through the points $x_0, \ x_1$, and x_2

$$x(t) = at^2 + bt + c, \tag{6.193a}$$

where the unknown values of the coefficients a, b, and c are determined as the result of solving a system of equations

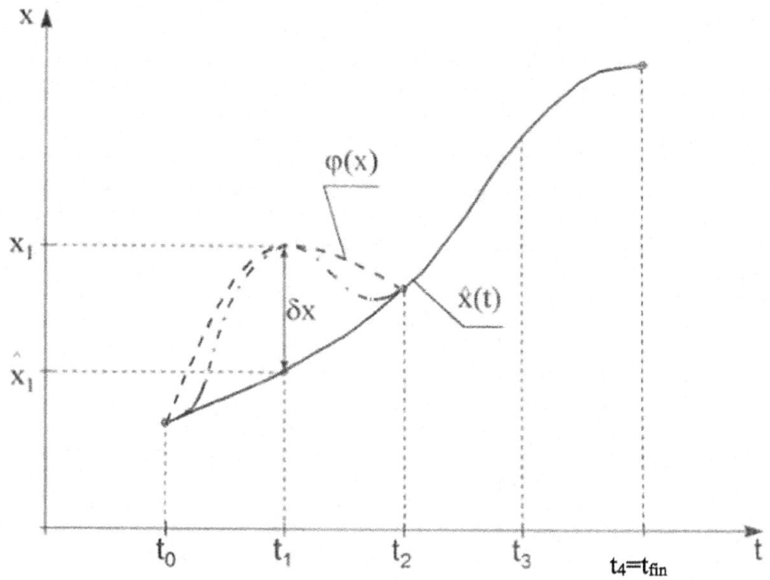

Fig. 6.58 Dividing the basic trajectory into arbitrary sections

$$\begin{bmatrix} t_0^2 & t_0 & 1 \\ t_1^2 & t_1 & 1 \\ t_2^2 & t_2 & 1 \end{bmatrix} \begin{bmatrix} a \\ b \\ c \end{bmatrix} = \begin{bmatrix} x_0 \\ x_1 \\ x_2 \end{bmatrix}, \tag{6.193b}$$

made on the basis of Eq. (6.193) for three points in time.

To determine the coordinates y and u in the point t_1 we successively differentiate the Eq. (6.193)

$$\begin{aligned} y(t_1) &= y_1 = \dot{x}_1 = 2at_1 + b; \\ u(t_1) &= u_1 = \ddot{x}_1 = 2a. \end{aligned} \tag{6.194}$$

Then, when the complete set of boundary conditions is obtained, the trajectory section is calculated on the interval $[t_0, t_1]$, and then on the interval a $[t_1, t_2]$, where x_1, y_1, u_1 act as final and initial conditions, respectively. In this case, the MDO algorithm works, according to the usual scheme, which provides the calculation of the section of the trajectory shown in Fig. 6.58 in the streak line which satisfies the initial and final conditions at the points t_0 and t_2 along coordinates x, y, and u.

At this part of the trajectory, the value of the functional ΔI_{0-2} is calculated and compared with the same component obtained for the basic trajectory $\Delta \hat{I}_{0-2}$. If $\Delta I_{0-2}<\Delta \hat{I}_{0-2}$ (minimization problem), then this section replaces the original one, and it becomes the basic one.

Then the variation δx is performed again, the coefficients a, b, c for the new value x_1 is calculated, y_1 and u_1 were determined and the new segment of trajectory is calculated. And so on, until.

$$\Delta\hat{I}_{0-2} - \Delta I_{0-2} \leq \varepsilon. \tag{6.195}$$

Next, the trajectory section is improved on the interval $[t_1, t_3]$ due to variation δx in the point t_2 with fixed ends (the new one on the left and the basic one on the right). Then the last section of the trajectory is improved on the interval $[t_2, t_{\text{fin}}]$.

This is the end of the first iteration. In the next iteration, you can try to improve the resulting trajectory in the considered way.

The described approach is easy to extend to the case of optimization of the trajectory of the AC movement in space [41–43].

Here, the most time-consuming task is to calculate the components of the vectors $y^T = [V\theta\varphi]$ and $U^T = \left[P\alpha\gamma\right]$ at the intermediate variable point.

We show, how these components are calculated. For this purpose, we write down in general form the equations of movement of the AC as a material point

$$\begin{aligned} \dot{L} &= f_1(V, \Theta, \varphi); \\ \dot{H} &= f_2(V, \Theta, \varphi); \\ \dot{Z} &= f_3(V, \Theta, \varphi); \end{aligned} \tag{6.196}$$

$$\begin{aligned} \dot{V} &= \Psi_1(P, L, \gamma, V, \Theta); \\ \dot{\Theta} &= \Psi_2(P, L, \gamma, V, \Theta); \\ \dot{\varphi} &= \Psi_3(P, L, \gamma, V, \Theta). \end{aligned} \tag{6.197}$$

By differentiating (6.196) and taking into account (6.197), we obtain a system of the second order for the subsequent calculation of the vector u at the variable point

$$\begin{aligned} \ddot{L} &= \eta_1(P, \alpha, \gamma, \Theta, \varphi); \\ \ddot{H} &= \eta_2(P, \alpha, \gamma, \Theta, \varphi); \\ \ddot{Z} &= \eta_3(P, \alpha, \gamma, \Theta, \varphi). \end{aligned} \tag{6.198}$$

To define the components of vectors x, y and u at the variable point, we will do the same as the example discussed above.

First, the variation L by the value of δL. is carried out. Then the coefficients a, b, and c of the parabola are calculated by the three points

$$L(t) = at^2 + bt + c \tag{6.199}$$

For the variable i-th point, using (6.194), the values of the derivatives $\dot{L}(t_i)$, $\ddot{L}(t_i)$ are calculated. The values of derivatives $\hat{\dot{H}}$, $\hat{\dot{Z}}$ and $\hat{\ddot{H}}$, $\hat{\ddot{Z}}$ is determined at the point t_i for the basic trajectory. The resulting values of the derivatives are substituted in the left parts of the expressions (6.196) and (6.198).

Let us solve the algebraic system (6.196), regarding unknowns V, θ, φ in the point t_i. Received values V_i, θ_i, φ_i substitute in (6.198), after which we can find the unknowns P_i, δ_i, γ_i. Thus, the vectors y_i and u_i in the point t_i are defined, and we can start calculating the section of trajectory. At the same time in the formulas of control (6.59) and (6.60), the required values of the derivatives are calculated as follows:

$$\frac{x_{\text{fin}} - x}{t_{\text{fin}} - t} = \left[\frac{L(t_i) - L(t)}{t_i - t}, \frac{\hat{H}(t + \Delta t) - H(t)}{\Delta t}, \frac{\hat{Z}(t + \Delta t) - Z(t)}{\Delta t}\right], \tag{6.200}$$

where the first element represents the required derivative to eliminate the discrepancy between the i-th and the current value of this coordinate, and the second and third elements provide the working out of basic values of height and lateral range.

Thus, the algorithm for calculating the trajectory works according to the usual scheme forms successively the sections of trajectory at intervals $\left[t_{i-1}, t_i\right]$ and $\left[t_i, t_{i+1}\right]$ with calculation of the functional $\Delta I_{i-1, i+1}$ and comparing it with the basic value for the same section $\Delta \hat{I}_{i-1, i+1}$. Next, the usual procedure is used to find the minimum of the functional as a function of three variables (L, H, Z).

At the same time, of course, we are talking about functions $L(t), H(t), Z(t)$ at interval $\left[t_{i-1}, t_{i+1}\right]$. Thus, as a result, we get the optimal section of the trajectory. Then the transition to the optimization of the next section in the interval $\left[t_i, t_{i+2}\right]$ etc. to the section $\left[t_{\text{fin}-2,}, t_{\text{fin}}\right]$ is carried out. At this point, the first iteration of obtaining the optimal trajectory ends. Taking this trajectory as a basic, we can make a successive improvement to its sections again. Experience shows that two or three iterations are enough to obtain an optimal trajectory corresponding to the exact solution.

As a recommendation for dividing the basic trajectory into sections, the following scheme should be followed.

In the beginning, the entire basic trajectory is divided in half in the sense of the time interval, i.e. $t_i = (t_{\text{fin}} - t_0)/2$.

After its improvement due to the variation of the coordinates of the state at the point t_i proceed to dividing in half of the intervals $[t_0, t_i]$ and $[t_i, t_{\text{fin}}]$ etc.

The advantage of this approach is that in the case of calculating the optimal trajectory, we always have «in stock» a realizable trajectory that is approached to the optimal one, and, if it satisfies the expected result of solving the applied problem and other submitted requirements, then the calculations can be ended.

In this case, the MDO differs from the principle of maximum in the most radical way during solving problems of the dynamics of the flight of the AC, because the worst case their solution for the MDO is a realizable trajectory that satisfies the boundary conditions and all restrictions on the coordinates of the state and control vectors. Obtaining a solution to these problems, using the principle of maximum, can be extremely difficult.

In conclusion of this subsection, we note that the use of the extended version of the MDO for solving test problems (item 6.4) gave the same results, but the computational costs in some examples increased from 2 to 5 times, compared to the original version of the MDO with a variety of derivatives. The reason for the increase in the volume of calculations is due to the trial variations, that were made regarding the basic solution «in a straight line».

Thus, the extended version of the MDO is a sequence of applying the usual MDO scheme «in a straight line» for the construction of a section of the trajectory, passing through some variable point in the neighborhood of the basic trajectory. The neighborhood itself is limited only by the conditions of satisfying the trajectory to the boundary conditions.

An extended version of the MDO, that operates with a difference of functions $\delta x(t) = \hat{x}(t) - x(t)$ at a given segment, in fact, it is a classical variational problem of minimizing the functional $I\,[x(t)]$, where $\delta x(t)$ is a variation (increment) of the argument $x(t)$ of the functional.

Therefore, the solution, obtained using the extended version of the MDO, satisfies the necessary condition of optimality

$$\frac{\partial I[x(t)]}{\partial x(t)} = 0. \tag{6.201}$$

Taking into account the possibilities of the MDO for implementing constraints, it can be assumed that its extended version occupies an intermediate position between direct variational methods and the principle of maximum in solving problems of the control of flight of the AC.

References

1. Alekseev VM, Tikhomirov VM, Fomin SV (1979) Optimal control. Nauka, Moscow
2. Afanasyev VN, Kolmanovsky VB, Nosov VR (1989) Mathematical theory of designing of the systems of control. Higher School, Moscow
3. Kozlov VI (1973) Lecture notes on the course. In: Variational methods as the basis design systems of automatic control system. MAI, Moscow
4. Lavrentiev MA, Lyusternik LA (1938) Course of variational calculus. Gonti NKTP USSR, M
5. Aleksandrov VV, Bakhvalov NS, Grigoriev KG et al (1988) Practicum on numerical methods in problems of the optimal control/Editorial Board. Moscow State University Press, Moscow
6. Elsholts LE (1969) Differential equations and calculus of variations. Nauka, The main editorial office of phys.-mat. Lit, Moscow
7. Atans M, Falb P (1968) Optimal control. Mashinostroenie, Moscow
8. Алексеев VM, Galeev EM, Tixomirov VM (1984) Collection of optimization tasks. Nauka, Moscow
9. Vasiliev FP (1980) Numerical methods for solving extreme problems. Nauka, Moscow
10. Marchuk GI (1980) Methods of computational mathematics. Nauka, Moscow
11. Richmeier R, Morton K (1972) Difference methods for solving boundary value problems. Mir, Moscow.
12. Krylov IA, Chernousko FL (1962) About the method of successive approximations for solving optimal control problems. Comput Math Math Phys 2(6):1132–1139
13. Amosov AA, Dubinsky YM, Kopchenova NV (1994) Computational methods for engineers. Higher School, Moscow
14. Bryson A (1972) Ho Yu-shi. Mir, Applied theory of optimal control. Moscow

15. Bellman R (1960) Dynamic programming. Foreign literature, Moscow
16. Belotserkovsky SM, Kochetkov YA, Krasovsky AA, Novitsky VV (1980) Introduction to aeroautoelasticity. Nauka, Moscow
17. Speedy K, Brown R, Goodwin J (1973) Theory of control. Mir, Moscow
18. Chucky F (1975) Modern theory of control. Nonlinear optimal and adaptive systems. Mir, Moscow
19. Taranenko VT, Momji VG (1986) Direct variational method in boundary value problems of flight dynamics. Mashinostroenie, Moscow
20. Akimov AN, Andreev VV (1997) The method of direct optimization. The main provisions and the order of application. Air Force Engineering Academy named by N.E. Zhukovsky, Moscow
21. Gill F, Murray W, Wright M (1985) Practical optimization. Mir, Moscow.
22. Gantmacher FR (1988) Matrix theory. Nauka, Moscow
23. Parkhomenko PP, Soghomonyan ES (1981) Fundamentals of technical diagnostics. Optimization of diagnostic algorithms, hardware tools. Energiya, Moscow
24. Dynamics of aircraft in the atmosphere. Terms, definitions, and designations. GOST 20058–80. Publishing House of Standards, M.1981.
25. Bertman AF, Aramanovich IG (1973) A short course of mathematical analysis for higher education institutions. Nauka, M.
26. Sokolov NI, Rutkovsky VYu, Sudzilovsky NB (1988) Adaptive systems of automatic control of aircraft. Mashinostroenie, M
27. Kondratenkov VA (1989) Control systems and onboard digital computing complexes of aircraft. Kiev
28. Samarskiy AA (1983) Theory of difference systems. Nauka, Main editorial office of phys.-mat. lit., M
29. Bakhvalov NS (1975) Numerical methods, vol 1. Analysis, algebra, ordinary differential equations. Nauka Main editorial office of phys.-mat. lit., M
30. Akimov AN, Andreev VV (1996) Numerical method of direct optimization in the tasks of optimal control. Theory Syst Control #3
31. Akimov AN, Vorobyev VV., Kibardin YA., Nikolaev YA (1989) Expert systems in the task of operational flight safety assurance / In Collection: Ensuring flight safety. Scientific and methodological materials // Ed. by S.A. Popitalov. -M.: Air Force Engineering Academy named by N.E. Zhukovsky, 1989.
32. Akimov A.N., Ryakin A.V. An approach to the construction of an onboard intelligent crew support system for the localization of active failures: A collection of materials of the 27th Interuniversity scientific and technical conference. Daugavpils, 1991.
33. Akimov A.N., Tretyakov A.V. Method of synthesis of optimal trajectories of movement of an aircraft with given initial and final conditions // In Collection: Ensuring flight safety. Scientific and methodological materials // Ed. by S.A. Popitalov. - M.: Air Force Engineering Academy named by N.E. Zhukovsky, 1992.
34. Akimov A.N., Tretyakov A.V., Ryakin A.V. Algorithms for constructing the optimal spatial trajectory of the aircraft movement. // In Collection: Ensuring flight safety. Scientific and methodological materials // Ed. by S.A. Popitalov. - M.: Air Force Engineering Academy named by N.E. Zhukovsky, 1992.
35. Ryakin AV. Algorithmic support of the onboard intelligent crew support system for faulure's localization // In Collection: Ensuring flight safety. Scientific and methodological materials // Ed. by V.A.Kondratenkov. - Kiev, 1992.
36. Ryakin AV (1992) Synthesis of the structure of the onboard intelligent crew support system for the localization of failure. In: Popitalov SA (Ed) Collection: ensuring flight safety. Scientific and methodological materials. Air Force Engineering Academy named by N.E. Zhukovsky, M.
37. Akimov A.N., Andreev V.V., Mohov A.A. The step-by-step increment method of single-stage aerospase plane three-dimension moving control functions determination. International Aerospase Congress, Theory, Applications, Technologies. Moscow, August 15–19, 1994, pp.11–17.

38. Momji VG, Taranenko VT (1985) Optimization of program movement of a nonlinear system with constraints. In book: Izvestiya Academy of Science USSR. Technical cybernetics # 2, pp 8–16
39. Calculation and analysis of the movement of aircraft. Engineering reference book. Mashinostroenie, Moscow 1985
40. Krutko PD (1987) Inverse problems of dynamics of controlled systems. Linear models. Nauka, Main editorial office of phys.-mat. lit., M
41. Zatuchny DA, Logvin AI, Nechaev EE (2012) Problems of implementing automatic dependent surveillance in the Russian Federation Monograph. MSTU CA
42. Zatuchny DA (2013) To optimize the route of the aircraft in the conditions of zonal navigation. Sci Bull MSTU CA (193):115–117
43. Eliseev BP, Kozlov AI, Romancheva NI, Shatrakov YG, Zatuchny DA, Zavalishin OI (2020) Probabilistic-statistical approaches to the prediction of aircraft navigation systems condition. Springer Aerospace Technology, 200p

The manufacturer's authorised representative in the EU is Springer Nature Customer Service Centre GmbH, Europaplatz 3, 69115 Heidelberg, Germany. If you have any concerns regarding our products, please contact ProductSafety@springernature.com

Printed and bound by CPI Group (UK) Ltd, Croydon, CR0 4YY
30/09/2026
02225308-0013